TRANSISTOR AUDIO AND RADIO CIRCUITS

Second Edition

The information in the first edition was based on material supplied by W. D. Benson, D. R. Hyde, and R. Osborne of Central Application Laboratory, Mullard Limited, and K. C. Warner formerly of Consumer Division, Mullard Limited. Additional material for the second edition has been supplied by J. A. Tijou and M. D. Brett of Central Application Laboratory. The second edition has been edited by M. J. Rose, Central Technical Services, Mullard Limited.

First edition 1969

Second edition 1972

ISBN 0 901232 39 4 TP1319

UDC 621. 375.4 : 621.396.62 : 621.382.3

Made and printed in England

by

WIGHTMAN AND COMPANY LIMITED, 1-3 Brixton Road, London, SW9 6DS

TRANSISTOR AUDIO AND RADIO CIRCUITS

Second Edition

Mullard Limited, Mullard House, Torrington Place,
London WC1 7HD

FOREWORD

The first edition of this book described itself as a manual of established and practical circuits for a wide range of equipment. The range extended from portable radio receivers to f.m. tuners and audio amplifiers capable of the highest standards of reception and reproduction. The same is true of the second edition.

New circuits have been added, reflecting developments that have occurred since the first edition was published in 1969. In particular, some integrated circuit designs have been included, and new high-quality amplifier and pre-amplifier circuits added. Most of the material of the first edition has been retained, and a new chapter on loudspeakers has been added.

The book is intended for use by equipment manufacturers, radio and television service engineers, university and college students, and the home constructor. The circuits presented can often be combined to produce a wide range of equipment. For example, many combinations are possible with the h.f. and a.f. stages for radiograms and portable radios. Again, with the high-quality circuits, the two f.m. tuners can be used with any of the audio amplifiers to give mono or stereo line-ups for use with disc and tape inputs as well.

Details of the practical circuits are complete, but because of the wide range of possible combinations of circuits, no attempt has been made to present precise layout and constructional information. Some general guidance is given for hi-fi amplifiers, and a chapter is devoted to some simple test equipment and its use in effectively commissioning and maintaining audio circuits. But because of the need for rather complex equipment for aligning r.f. and i.f. circuits, similar information has not been given for such stages: it is envisaged that it will be those who have access to specialised knowledge and equipment—amateur enthusiasts, beginners under expert tutelage, or professionals—who attempt to build these h.f. circuits.

The circuits in this book are proven designs, and prototypes have been built from which the performance figures quoted were derived. It is emphasised, however, that Mullard Limited cannot undertake to answer individual queries or resolve problems encountered by individual constructors.

CONTENTS

AVAILABILITY OF COMPONENTS

All the Mullard devices mentioned in this book can be obtained through the usual trade channels*. Other components, including meters and transformers, not marketed by Mullard are also readily obtainable. The addresses of manufacturers of components mentioned in the text are given below.

Transformers

Colne Electric Limited, Bury Lane, Rickmansworth, Herts.

Drake Transformers Limited, Kennel Lane, Billericay, Essex.

Gardners Transformers Limited, Summerford, Christchurch, Hants.

Parmeko Limited, Percy Road, Aylestone Park, Leicester.

Meters

Sifam Electrical Instrument Company Limited, Woodland Road, Torquay, Devon.

Taylor Electrical Instruments Limited, Archcliffe Road, Dover, Kent.

Printed-wiring Boards.

Printed-wiring boards for some of the audio circuits described in this book may be obtained from:

Bribond Printed Circuits Limited, Terminus Road, Chichester, Sussex.

Hart Electronics Limited, Penylan Mill, Morda, Oswestry, Salop.

DATA ON MULLARD DEVICES

Abridged data on most of the Mullard devices mentioned in this book are given in the current edition of the Mullard 'Pocket' Data Book (U.K. edition) available from technical booksellers.

Complete data on all current Mullard devices are given in the Mullard Technical Handbook. Details of the latest edition of the Handbook, and data sheets on the few devices not included in the 'Pocket' Data Book, can be obtained from:

Central Technical Services, Mullard Limited,
New Road,
Mitcham,
Surrey, CR4 4XY.

---

*See note on electrolytic capacitors, page 262

CHAPTER 1

SILICON AND GERMANIUM TRANSISTORS

Most transistors are made from either silicon or germanium. The early transistors were made from germanium because silicon presented many manufacturing problems, but these problems have been largely overcome by the development of the planar process. Silicon and germanium transistors are now produced concurrently, and there are applications for which each is the more suitable.

Certain characteristics of germanium semiconductor material limit the operating junction temperature of germanium transistors to below 100°C, while silicon devices can operate at temperatures in excess of 150°C. However, germanium transistors still have the advantage of a low value of forward base-emitter voltage and bottoming voltage and, since these two represent voltage losses in the circuit, germanium devices are particularly suitable for the lower-voltage type of a.f. output stage. Another advantage of germanium transistors is that they are easily matched in complementary (p-n-p/n-p-n) pairs, thus facilitating their use in complementary-pair push-pull output stages.

Because silicon has an extremely high intrinsic resistivity the leakage currents of silicon transistors are far smaller than those of germanium transistors. This advantage, together with the higher permissible junction temperature mentioned above, and the high voltage ratings available with silicon transistors, makes them ideal for a wide variety of audio applications, especially at high output powers where high voltages are necessary. Silicon planar transistors can be manufactured with high switching speeds, making these devices suitable for use at high frequencies, for example in the r.f. stages of radio receivers.

TRANSISTOR MANUFACTURE

The manufacturing processes of germanium and silicon transistors both start with the production of a crystal of the doped semiconductor material. The crystal is then cut into thin slices and it is from this point on that the manufacturing processes differ.

Germanium Transistors

Germanium transistors can be made by either of two methods—the alloy-junction method or the alloy-diffusion method. In the manufacture of a p-n-p transistor by the alloy-junction method, an emitter pellet of p-type material is fused to one side of the n-type slice and a collector pellet of p-type material is fused to the other side. In the alloy-diffusion method, a p-n-p transistor is made by depositing a layer of n-type material on one surface of the p-type slice and placing two pellets close together on the n-type layer. One of the pellets, which eventually forms the base connection of the transistor, contains n-type additives only, and the other pellet, which will form the emitter, contains p-type and n-type additives. The assembly is heated and the additives diffuse into the p-type germanium beyond the prediffusion layer. In this way a transistor is made in which the collector is formed by the original slice, and all diffusion has been effected on one side. The processes used in the manufacture of germanium transistors are now well known and will not be elaborated upon in the present book. The manufacture of a germanium alloy-junction transistor is summarised in Fig. 1.

Silicon Transistors

A process now widely used in the manufacture of silicon transistors is the planar process illustrated in Fig. 2. A slice of n-type silicon is cleaned under very stringent conditions and is then heated in oxygen so that a layer of silicon oxide is formed all over it. The oxidised slice is

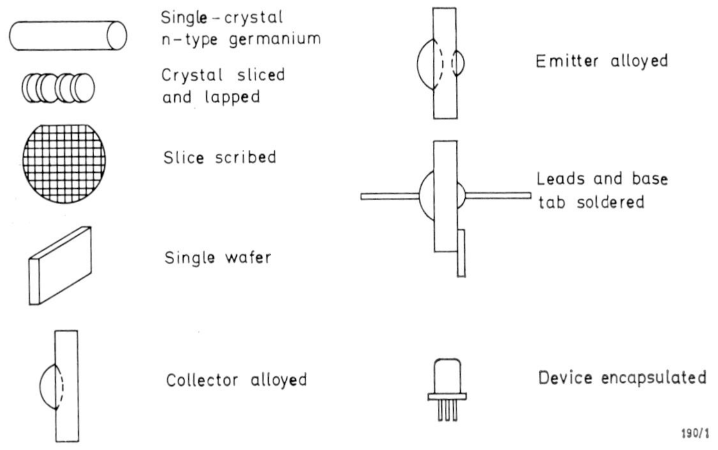

Fig. 1—Stages in the manufacture of germanium alloy-junction transistors

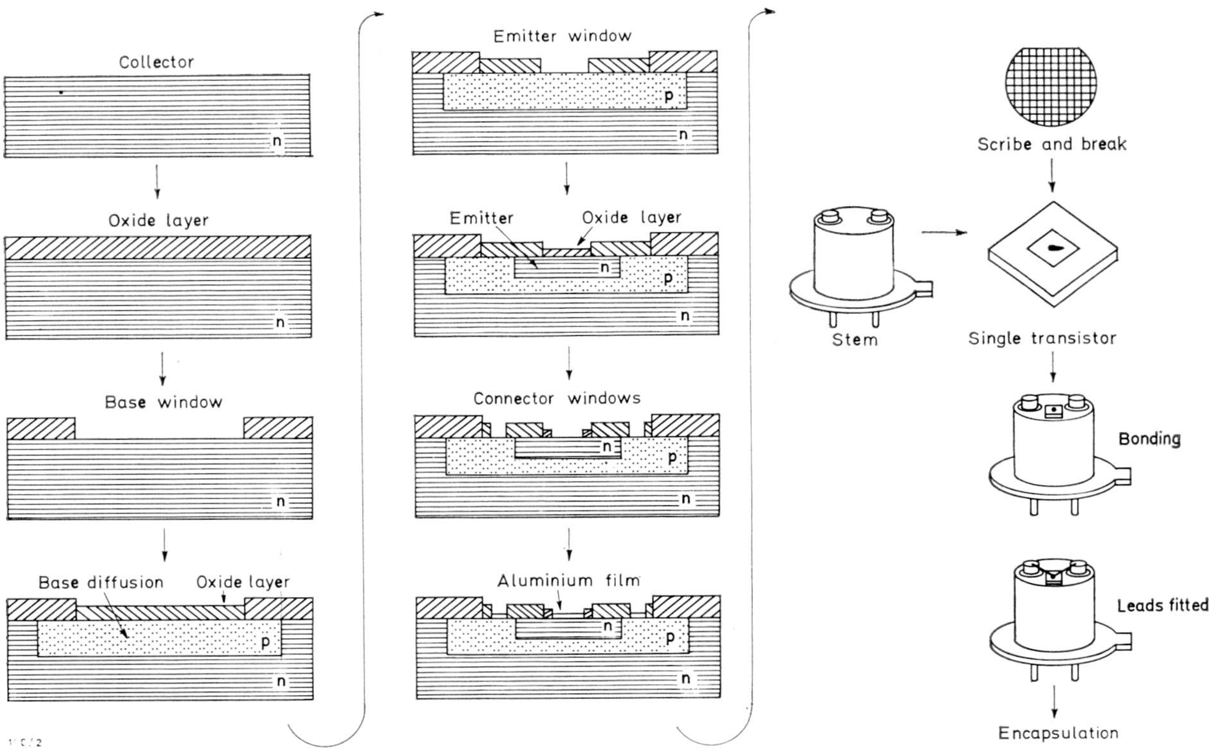

Fig. 2—Stages in the manufacture of silicon planar transistors

attached to the middle of a high-speed spinner, and a drop of photosensitive etchant-resistant material, or photoresist, is placed on it. When the spinner revolves, the photoresist spreads out in a thin, even coating. Some transistors, such as power transistors, need a thicker layer than others; in high-frequency transistors where well-defined edges are needed, a thin layer of photoresist is used. Hence, the spinning speed for power transistors is less than that for high-frequency transistors.

Base and Emitter Diffusion

When the slice has been coated with the photoresist, it is covered with the base mask and exposed to ultraviolet light. The base mask is a photographic negative which carries a pattern formed by the proposed positions of the bases of the transistors that will be made from the slice.

The parts of the photoresist exposed to ultraviolet light change to a hard material which cannot easily be washed off, while the unexposed parts are easily removed. Therefore, after washing, there are many small windows in the photoresist, through which the silicon oxide is exposed. These windows are in the positions of the transistor bases.

The remaining photoresist is further hardened by baking the slice, which is then treated with acid to etch away the oxide exposed through the windows and reveal the silicon. The rest of the photoresist is then removed and the slice is ready for base diffusion.

When all the photoresist has been removed, the slice is again cleaned and placed in an oven with a little of the required p-type doping additive (usually boron). The additive evaporates and some of it settles in the windows. The slice is then transferred to a hotter oven filled with oxygen. In this oven, the additive is driven into the silicon to the required depth. At the same time, another oxide layer is formed all over the slice, including the windows. The base of each transistor in the slice has now been formed.

The slice is again covered with the photoresist and another pattern of windows produced as described above. This time, however, a different negative—the emitter mask—is used, and the windows are in the positions of the transistor emitters. The cleaning, diffusion and oxidation processes are repeated to form the transistor emitters. The doping additive used in this diffusion process, however, is not boron but phosphorus—an n-type substance.

Lead Connections

By means of the photoresist and a third negative, windows are produced over the base and emitter areas. The slice is placed in a vacuum and a fine layer of aluminium deposited all over it. The slice is removed from the vacuum and, by means of a fourth mask and another etching process, the aluminium is removed from all areas except where it is needed to provide contacts to the bases and emitters. The remaining aluminium is lightly alloyed to the bases and emitters to form non-rectifying contacts. At a later stage, fine gold or aluminium wires will connect the aluminium layer to the transistor leads or pins.

The oxide layer is now removed from the collector side of the slice, and the thickness of the slice is reduced by etching to lower the thermal resistance and improve the knee characteristic of the transistors.

The slice now contains between 200 and 6000 transistors, depending on the type of transistor, to which no leads have been attached. Each transistor is tested for breakdown voltage and small-signal current gain by means of a special probing machine connected to a test set. A photograph of this machine is given in Fig. 3. Transistors which fail any of these tests are marked and rejected at a later stage in manufacture.

Fig. 3—Probing machine checking breakdown voltage and current gain of transistors

After these tests, the slice is scribed with a diamond and broken into dice, each containing one transistor. At this stage, the marked transistors are rejected.

Each good transistor is soldered or bonded, collector side down, on the gold-plated stem or header. This operation is performed automatically on a dice bonder which fixes the transistor to the header by a silicon-gold eutectic.

The wires connecting the base and emitter to their leadout wires are usually of gold or aluminium, and are attached to the aluminium film on the transistor by means of a suitable bonder. When the leads have been attached, the transistors are encapsulated.

Lockfit Encapsulation

The familiar metal-can encapsulation requires no further description. However, a more modern encapsulation is epoxy resin, which is used on the range of Lockfit transistors. The coating of epoxy material provides the semiconductor wafer with complete environmental protection, and the junction-to-ambient thermal conductivity of the material is superior to that obtained with most metal encapsulations. The transistor leads take the form of flat pins which are designed to snap into a printed-wiring board as shown in Fig. 4.

Fig. 4—Insertion of Lockfit transistors in printed-wiring board

The encapsulated transistors are subjected to a temperature cycling test, a drop test and a test to detect pinholes in the encapsulation.

Epitaxy

When special types of silicon planar transistor are required, the procedure is modified slightly. When, for example, transistors are required for r.f. and i.f. applications, the silicon substrate is more heavily doped than in other transistors, and it undergoes further treatment in the crystal-growing chamber before being cleaned and covered with the first coat of photoresist. Consequently, the slice is covered with a thin layer—5 to 20 micrometres thick—of lightly-doped silicon. This layer has the same crystal orientation as the original slice and is therefore known as the epitaxial layer. The subsequent treatment of the slice with the epitaxial layer is the same as previously described. In the epitaxial type, however, the transistor is essentially within the thin layer, and the substrate is really a low-resistance back contact which provides a convenient thickness for handling. With the non-epitaxial type, the transistor is within the substrate.

Field-effect Transistors

Despite the differences in materials and methods of manufacture, the transistors described in the earlier sections of this chapter all have essentially the same construction and principle of operation. They all consist of emitter, base and collector regions, and their properties are all largely a consequence of the movement of electrons and holes across the junctions between these regions or of the existence of the junctions themselves. The field-effect transistor, however, has a structure and characteristics differing from the conventional junction transistor.

Structure and Operation

Fig. 5 is a schematic diagram of a field-effect transistor. It will be seen that a channel of n-type semiconductor material is enclosed for part of its length by a collar of p-type material. One end of the n-type channel is called the source, and the other end is called the drain. The collar of p-type material is the gate.

If a voltage is applied across the length of the n-channel so that the source is positive with respect to the drain, electrons will flow from the drain to the source; that is, current will flow conventionally from the source to the drain.

If a voltage is applied to the gate so that the gate is negative with respect to the source, the p-n junction formed between the gate and channel will

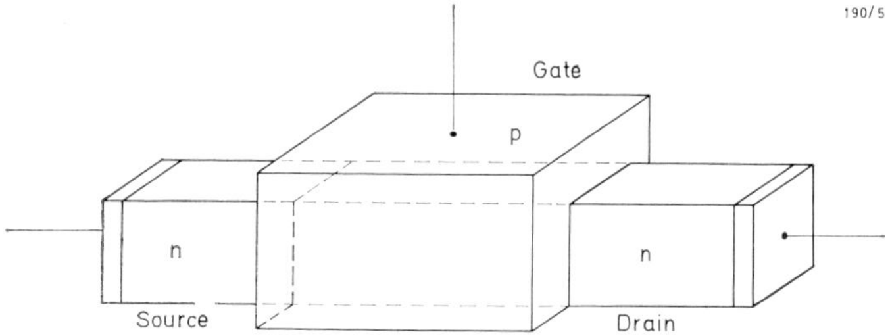

Fig. 5—Schematic diagram of a field-effect transistor

be reverse-biased. No current flows across the junction, but the recombination of holes and electrons that migrate across the junction forms a region in the n-channel depleted of free electrons.

This non-conductive depletion region narrows the conductive n-channel, and thus reduces the current flowing from source to drain. Varying the gate bias voltage will thus vary the source-drain current. The action is analogous to the control of anode-cathode current in a triode valve by means of a negative bias voltage applied to the grid of the valve.

In the field-effect transistor, as in the triode valve, a comparatively small bias voltage will control a large current. Therefore amplification of a signal is possible with the field-effect transistor.

Since the gate-channel junction is reverse-biased during operation, the field-effect transistor is essentially voltage driven, and it therefore has a high input impedance. The device has a low noise figure and good high-frequency performance. For these reasons, field-effect transistors are ideal for use in such applications as the r.f. stages of high-quality f.m. tuners.

Manufacture

The structure of an n-channel field-effect transistor is shown in Fig. 6. The planar process is employed in the manufacture of these transistors.

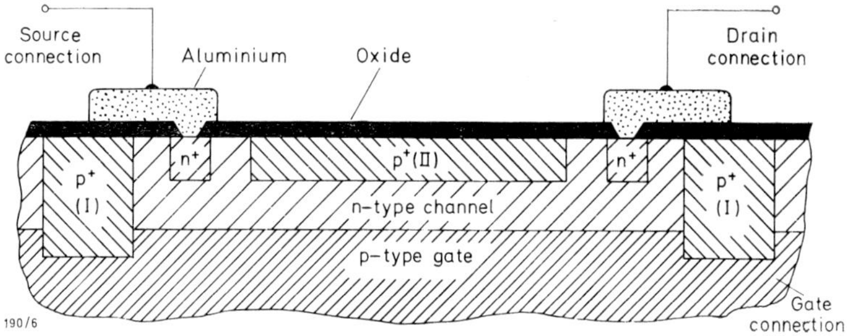

Fig. 6—Structure of a field-effect transistor

An n-type epitaxial layer which eventually will form the n-channel is grown on a substrate of high-conductivity p-type material, which will form the gate. After oxidation of the surface of the epitaxial layer, windows are etched in the oxide layer corresponding to the channel-isolating region I. Through these windows is diffused an additive rich in holes, which creates a 'super-p' or p^+ region penetrating through the epitaxial layer to the p-type substrate, and thus isolating the n-channel.

After re-oxidation, a window is etched corresponding to region II, and another p^+ region is created to form the upper part of the p-gate surrounding the n-channel.

A third diffusion is performed to create 'super-n' or n^+ regions at the ends of the n-channel. Aluminium bonds are deposited on these n^+ regions, and to these are connected the source and drain contact wires. The purpose of the n^+ region is to diminish the series resistance of the source and drain contacts. The gate contact is made to the p-type substrate, and the transistor is encapsulated in a metal envelope.

CHAPTER 2

BASIC H.F. CIRCUITS

The block schematic diagram for the high-frequency stages of a typical a.m./f.m. radio receiver is given in Fig. 7. The audio-frequency, or a.f., output is fed to one of several types of a.f. amplifier, which are discussed more fully in Chapter 3. In the present chapter, the operation of each of the circuits named on the block diagram is discussed in general terms, detailed practical circuits being given in later chapters.

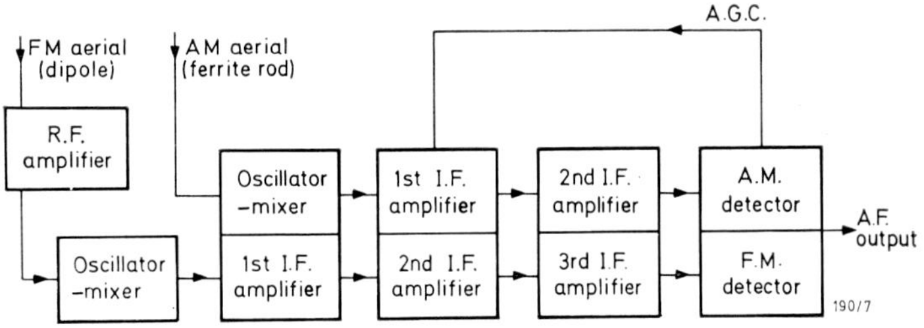

Fig. 7—H.F. stages of a typical a.m./f.m. receiver

R.F. AMPLIFIER

A typical basic common-emitter r.f. amplifier circuit is shown in Fig. 8. For f.m. reception, the transistor is often connected in the common-base configuration because this configuration gives the highest gain at frequencies approaching the cut-off frequency. The collector circuit is tuned and is inductively coupled to the self-oscillating mixer circuit.

MIXER

The mixer, or frequency changer, converts by heterodyne or 'beating' action the wanted frequency of the carrier signal at the aerial to a fixed frequency at the input to the i.f. amplifier. The a.m. carrier frequencies vary from 160 to 280kHz in the long waveband and from 540 to 1640kHz in the medium waveband. The f.m. carrier frequencies vary from 85 to 100MHz. The intermediate frequency is usually 470kHz for a.m. and 10·7MHz for f.m. The audio-frequency information, carried on the r.f. waveform as modulation of the amplitude or frequency, is transferred from the r.f. carrier signal to the i.f. carrier.

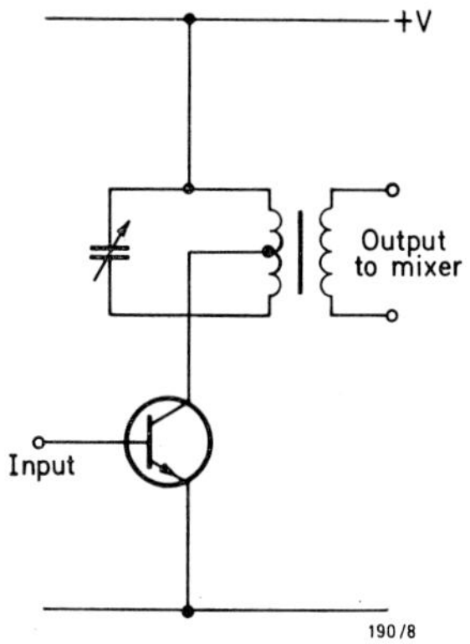

Fig. 8—Basic r.f. amplifier circuit

Heterodyne action is effected by injecting signals of different frequencies into a non-linear element such as the base-emitter diode of a transistor. In a radio receiver, one injected signal is the aerial signal, derived either directly or through an r.f. amplifier, and the other is a signal generated within the receiver by a local oscillator. The signals combine as shown in Fig. 9 to produce a complex waveform. When the waveform is applied between the base and emitter of a transistor, the main components occurring in the collector waveform are the original aerial and oscillator frequencies together with their sum and difference frequencies. The difference frequency (i.f. signal) is preferentially selected by the tuned circuit in the collector of the mixer, and further selectivity is provided by the i.f. amplifier stages. Voltage gain is obtained from the mixer stage in the same manner as the gain obtained from a normal i.f. amplifier, but somewhat reduced in magnitude.

To maintain a constant difference between the aerial and oscillator frequencies, the oscillator frequency must be changed whenever the aerial frequency is changed, that is, the oscillator tuned circuit must 'track' the aerial tuned circuit. Therefore, if the aerial is tuned to 1·5MHz, the local oscillator must be tuned to 1·97MHz, to give a difference frequency of 470kHz, and if the aerial tuning is changed to 1MHz, the oscillator must be tuned to 1·47MHz. For a.m. reception, the oscillator is usually tuned to frequencies 470kHz above the aerial frequency so that signals in the long waveband, which have frequencies below 470kHz, can be accepted. For f.m. reception the oscillator frequency may be 10·7MHz above or below the aerial frequency dependent upon other design considerations.

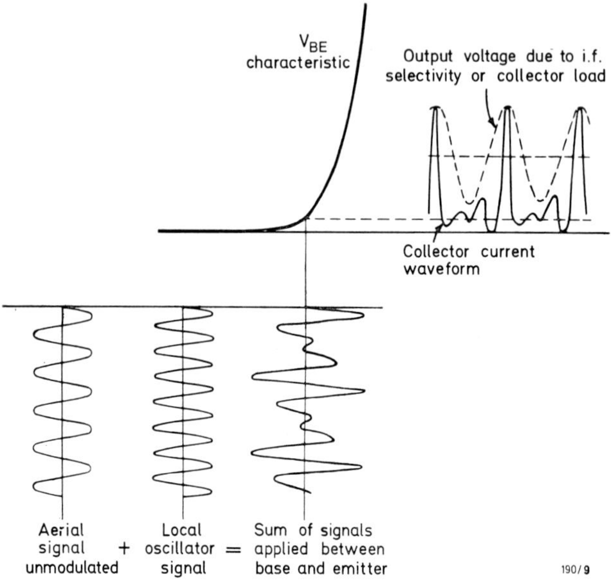

Fig. 9—Diagram illustrating the action of a mixer transistor

SELF-OSCILLATING MIXER

The local-oscillator signal applied to the mixer may be derived from a separate oscillator transistor in the receiver or one transistor may act as both oscillator and mixer. This latter arrangement is the more common and is known as a self-oscillating mixer.

The basic circuit of a self-oscillating mixer is shown in Fig. 10. At the oscillator frequency, the base of the transistor is effectively connected to earth because the aerial tuned circuit has negligible impedance. The input electrode for the oscillator action is, therefore, the emitter, which is inductively coupled to the collector of the transistor. Both the emitter winding and the collector winding are closely coupled to the oscillator tuned circuits. Positive feedback is applied from output to input through this coupling, and oscillation occurs at the resonant frequency of the oscillator tuned circuit. The aerial signal is applied from the aerial tuned circuit to the base of the transistor. Both the aerial signal and the oscillator signal are, therefore, in series with the base-emitter diode. The component

Fig. 10—Basic self-oscillating mixer circuit

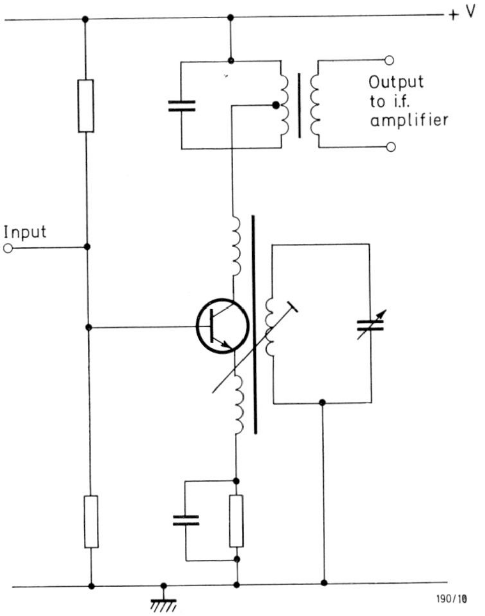

of the heterodyne output signal having the required intermediate frequency is extracted by the i.f. bandpass filter or transformer in the collector circuit.

A typical self-oscillating mixer for a.m. reception is shown in Fig. 11. The circuit is shown for one waveband only, the different wavebands being obtained by switching other coils into the aerial circuit and additional capacitors into the oscillator circuits. The aerial input is taken to the base of the transistor, and the oscillator feedback from the collector is taken through low-impedance coupling coils to the emitter.

To ensure that the oscillation starts easily, the transistor is biased initially for class A operation by the potential-divider bias circuit. As the oscillation increases in amplitude, rectification of the oscillating signal occurs at the base-emitter diode. This causes the loop gain after each cycle to decrease, stabilising the amplitude of the oscillation and causing the quiescent emitter current to rise slightly (from 250 to 300μA in the circuit shown in Fig. 11).

Correct tracking of the aerial and oscillator tuned circuits to maintain the 470kHz difference in frequency can be obtained by a conventional padder circuit or by using a tuning capacitor with specially-shaped vanes for the oscillator circuit. The value of the tuning capacitance is not critical but must be sufficient to provide the required frequency coverage.

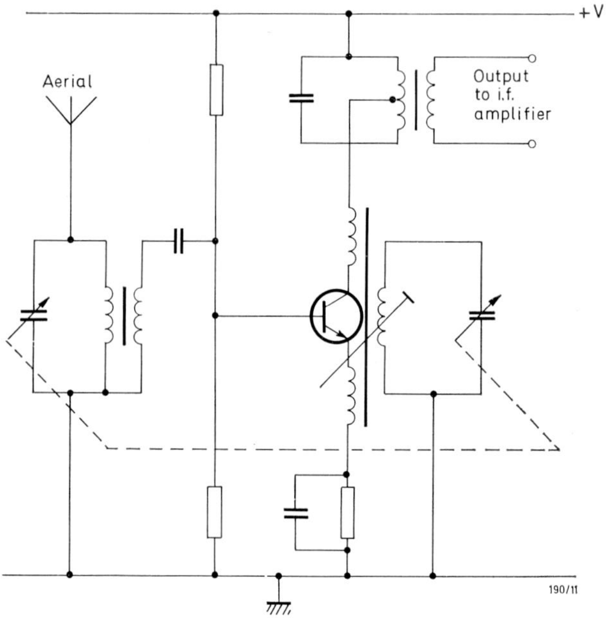

Fig. 11—Typical practical self-oscillating mixer circuit

Stray capacitance can exist between the aerial and oscillator sections of the tuning capacitor and so form a path for unwanted feedback between the two sections. To prevent this feedback and the resulting possibility of spurious oscillations, a screen is usually placed between the two sections. Another source of feedback between the oscillator and aerial sections is through the stray capacitance of the leads to the wave-change switch, and these leads should, therefore, be kept as short as possible.

I.F. AMPLIFIER

The basic circuit of an i.f. amplifier stage is shown in Fig. 12. The input and output transformers are tuned to the intermediate frequency, the selectivity and bandwidth being improved if both windings of the transformers are tuned. In an a.m./f.m. receiver it is necessary for two i.f. transformers to be provided between stages—one tuned to 470kHz and the other to 10·7MHz. A method of connecting these i.f. transformers is illustrated in Fig. 13. The remaining i.f. stages are similar to the first except that, in an a.m./f.m. receiver, the a.m. and f.m. outputs must be separated so that each can be fed to its own detector circuit.

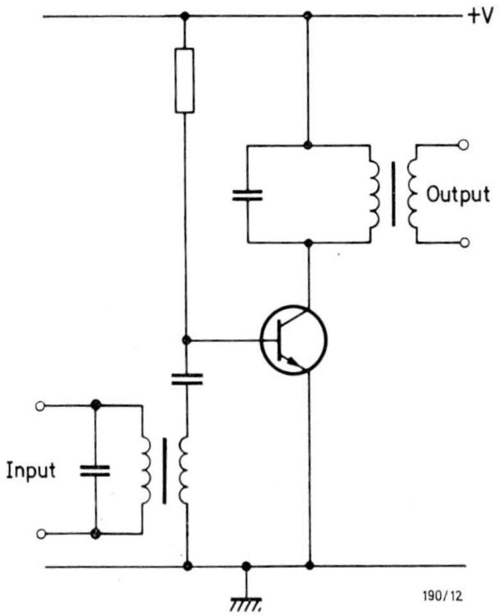

Fig. 12—Basic i.f. amplifier circuit

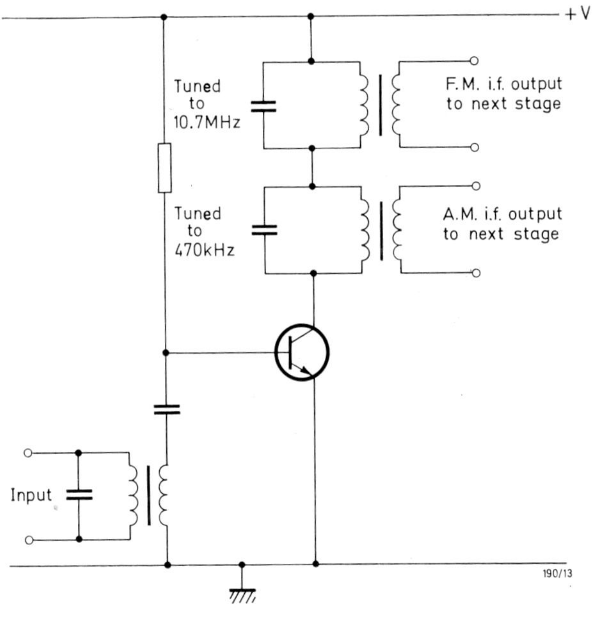

Fig. 13—Basic i.f. stage for an a.m./f.m. receiver

DETECTOR

The a.m. detector of a transistor receiver is usually a semiconductor diode feeding a potentiometer load. The potentiometer forms the volume control, which feeds the demodulated signal to the audio-frequency stages of the receiver. The potentiometer is usually shunted by a 10nF capacitor, as shown in Fig. 14, to filter out the i.f.

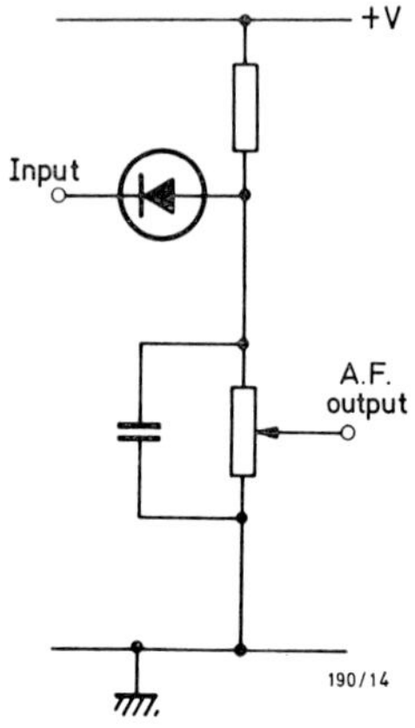

Fig. 14—Basic a.m. detector circuit

The detector stage for f.m. signals is usually of the type known as a ratio detector, the basic circuit of which is shown in Fig. 15.

The ratio detector depends for its operation on the fact that as the frequency of the signal varies, the phase relationships in a circuit tuned to the nominal frequency change. The transformer has three windings, the primary and secondary tuned to the intermediate frequency and the tertiary, close-coupled to the primary, injecting a voltage into the secondary

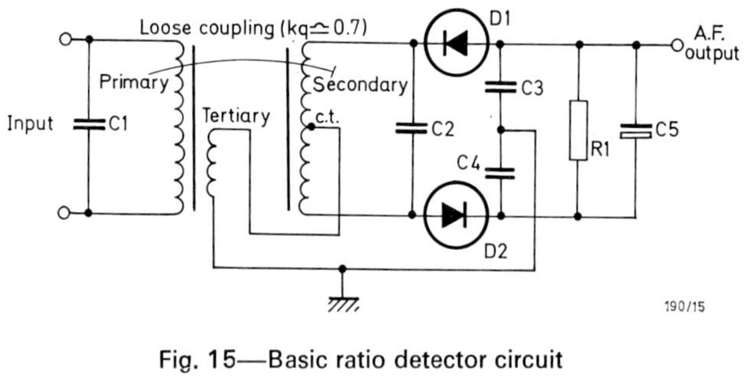

Fig. 15—Basic ratio detector circuit

in series with the voltage across each half of the secondary. Figs. 16 and 17 show the phase relationships. In Fig. 16, the voltage V_S induced in each half of the secondary is 90° out of phase with the injected primary voltage V_P at the intermediate frequency. Voltages V_{D1} and V_{D2}, applied to the detector diodes, are equal in magnitude, so there is no audio output.

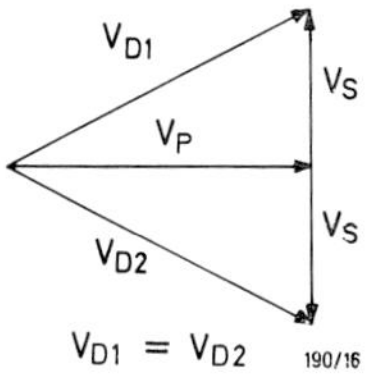

Fig. 16—Vector diagram of voltages in a ratio detector circuit with signal at nominal frequency

Fig. 17 shows the phase conditions when the frequency of the signal differs from the nominal value. The effect of a shift in frequency is to change the phase of the secondary voltage with respect to the injected primary voltage. As the frequency of the signal sweeps above and below the nominal value the variation of the relative phase of the injected primary and secondary voltages produces a variation in the amplitudes of V_{D1} and V_{D2} at audio frequency. The voltages are rectified by the two diodes and the a.f. signal taken from across either C_3 or C_4. The long-time-constant circuit R_1C_5 suppresses any unwanted a.m. signals.

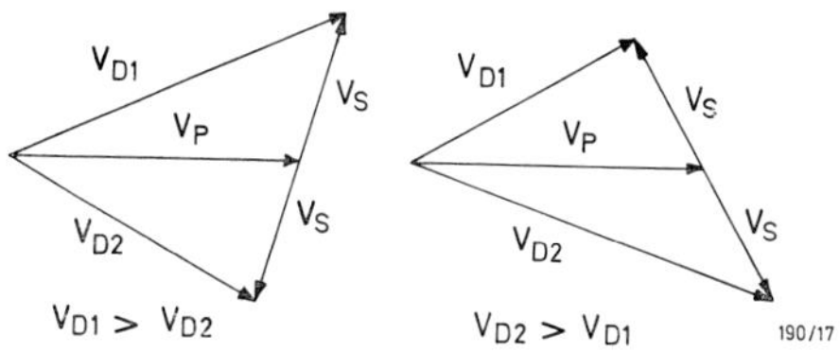

Fig. 17—Vector diagram of voltages in a ratio detector circuit with signal varying from nominal frequency

AUTOMATIC GAIN CONTROL

It is necessary for the sensitivity of r.f. and i.f. stages in a receiver to be high, so that weak signals are amplified to give a useful output. However, if no automatic control of gain were incorporated in an a.m. receiver, clipping would occur when the receiver was tuned to a strong signal. The automatic gain control (a.g.c.) signal, proportional to the i.f. output, is derived from the detector and applied in series with the base-emitter voltage of the first i.f. transistor (and the r.f. transistor in car radios). The a.g.c. signal therefore controls the bias, and hence the gain, of the transistor. The simplest method of modifying the basic detector and i.f. amplifier is shown in Fig. 18.

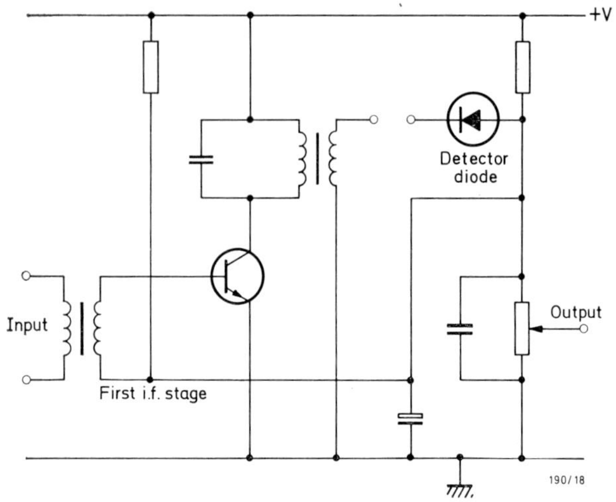

Fig. 18—Simple a.g.c. circuit

To eliminate the audio- and radio-frequency signals from the a.g.c. voltage, the decoupling capacitor of the potential divider feeding the base of the first i.f. transistor is large—8μF is a typical value. The a.g.c. characteristic may be improved by adding a damping diode across the first i.f. transformer. This diode ensures that the a.g.c. action is effective over a larger range of r.f. input voltages than in a normal circuit.

In f.m. receivers the waveform is already limited for a.m. suppression and a.g.c. is generally thought to be unnecessary, although it may be included to improve the performance of the i.f. stages.

BIAS AND STABILISATION AGAINST TEMPERATURE VARIATIONS

In the foregoing discussion of r.f. and i.f. circuits, no mention is made of biasing requirements. The operating current of a transistor is determined by the biasing voltages or currents applied to it, and these are usually set by means of resistors in the base, emitter and collector circuits. Because of the superior high-frequency performance of silicon transistors, germanium devices are not used in any of the r.f. and i.f. stages described in this book. Therefore only biasing arrangements for silicon transistors in these stages are described in detail.

Ideally, biasing circuits should allow a particular value of operating current to be obtained irrespective of changes in supply voltage and temperature; and spreads in the current amplification factor, h_{FE}, and the base-emitter voltage, V_{BE}, of the transistor. Unless an external stabilising element is used, it is impossible to eliminate the effects of variations in supply voltage; but in a good biasing arrangement, the resulting variations in current will not be much greater than the percentage variations in the supply voltage.

Different biasing arrangements satisfy these requirements to different degrees. The biasing arrangement shown in Fig. 19 has been almost universally adopted for low-level class A amplifiers using germanium transistors. In this arrangement the current is only slightly influenced by spreads in h_{FE}, but a value of emitter voltage greatly in excess of V_{BE} is required if the current variations resulting from supply voltage variations are to be small. On the other hand, a single biasing resistor in the base circuit, as shown in Fig. 20, gives negligible changes in current with V_{BE} variations, but the current in this circuit is fully dependent on spreads in h_{FE}.

The two major differences between the characteristics of silicon and

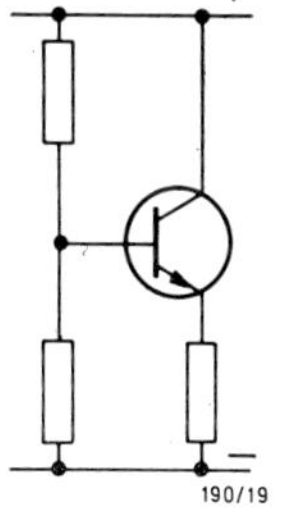

Fig. 19—Conventional voltage-biased transistor

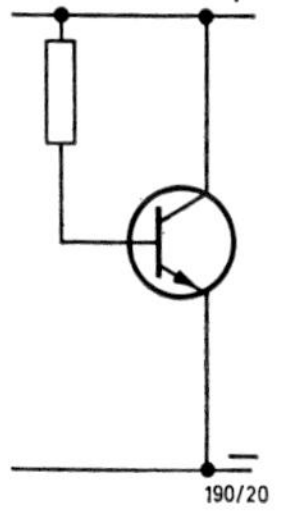

Fig. 20—Conventional current-biased transistor

germanium transistors that affect the biasing arrangements occur in the base-emitter voltage V_{BE} and collector leakage current I_{CBO}. The base-emitter voltage of silicon transistors at low currents is 0·6 to 0·7V, and of germanium transistors, 0·1 to 0·2V; the typical mutual characteristics of a silicon and a germanium transistor are compared in Fig. 21. The

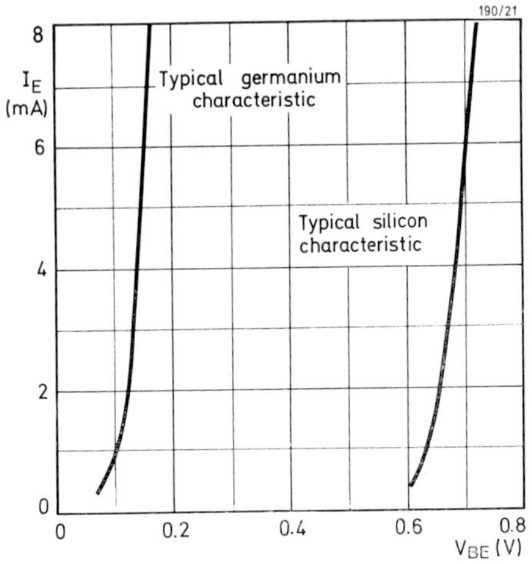

Fig. 21—Typical mutual characteristics of a germanium and a silicon transistor

difference in V_{BE} is equivalent to a difference in 'starting' voltage, but the rate of change of current with base-emitter voltage above this 'starting' voltage is identical for silicon and germanium transistors. The only consequence of this difference in V_{BE} is that, for conventional voltage biasing shown in Fig. 19, a larger drop across the emitter resistor will be required with silicon transistors than with germanium transistors for comparable changes in operating current with supply voltage, unless a stabilised source of bias voltage is available.

The collector leakage current in silicon transistors is negligible up to the highest temperatures likely to be encountered in radio equipment. Consequently, it would seem to be possible to use a high value of d.c. base source resistance, and to dispense with the normal stabilising circuit, which applies negative d.c. feedback to stabilise the operating conditions. However, other effects need consideration and these may necessitate retaining the stabilising circuit and limiting the magnitude of the source resistance. These effects are variations in supply voltage, spreads in V_{BE} and

h_{FE}, and variations in V_{BE} and, to a lesser extent, h_{FE} with temperature.

The three most common biasing arrangements are voltage biasing with emitter feedback, current biasing with collector feedback, and simple current biasing. The choice of biasing arrangement depends very much on the supply voltage, the proportion of this voltage that can be dropped across a feedback resistor, and the type of transistor used. A comparison of the spread in emitter operating current resulting from spreads in h_{FE} and V_{BE}, for each of the three biasing arrangements, is given in Table 1. The changes in operating current when the supply voltage falls to half its nominal value are also compared. An analysis of the biasing arrangements, and worked examples showing their relative effectiveness, are given in Appendix 1.

TABLE 1

Comparison of Biasing Arrangements

Nominal values of parameters:

$V_S=7V$; $V_{BE}=0{\cdot}7V$; $h_{FE}=100$
$I_E=1mA$ (approx.)

Biasing arrangement	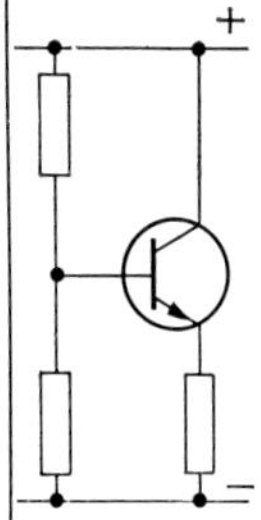	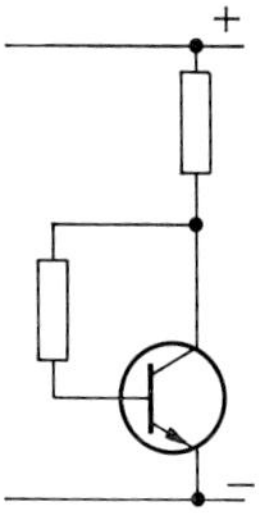	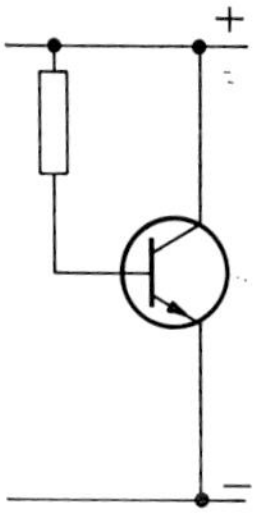
Parameter variation		I_E variation	
h_{FE} ±50%	+1, -1·6%	+23, -35%	±49%
V_{BE} ±7%	±1·7%	±0·95%	±0·86%
V_S −50%	−62%	−55%	−56%

CHAPTER 3

BASIC A.F. CIRCUITS

Audio-frequency amplifiers range from the simple amplifier used in small portable radio receivers, where size and cost are the important factors, to the large multi-stage circuit where cost is of little importance but reproduction of the highest quality is required. In the present chapter, only the basic amplifying stages are considered, the refinements necessary for high-quality reproduction being considered in Chapter 7.

The only class of one-transistor amplifier stage in common use is class A. Push-pull stages, either using two similar transistors or a matched complementary pair, may operate in class AB or class B. A special type of class AB push-pull amplifier sometimes encountered is known as 'π-mode'.

CLASS A OPERATION

Class A operation is that in which the values of bias and signal voltage applied to the transistor ensure that collector current always flows. Class A output stages are used mostly in car radios, where current drain is relatively unimportant, although class A circuits are invariably used as the earlier stages of multi-stage amplifiers where the current level is comparatively low. Class A, one-transistor output amplifiers are inefficient and have the additional disadvantage of suffering from even as well as odd harmonic distortion, although they are simple and enable moderate powers to be obtained from circuits consisting of very few components.

A basic class A output stage is shown in Fig. 22. The bias is provided by resistors R_1 and R_2, and resistor R_3 provides thermal stabilisation.

PUSH-PULL OPERATION

Push-pull output stages are used almost universally in modern amplifier circuits, and push-pull circuits are often used in the driver stages of multi-stage amplifiers. The output power obtained from a pair of transistors connected in a push-pull configuration is considerably more than double the power obtained from one transistor. Push-pull operation also gives rise to less of the even harmonics sometimes associated with single-transistor, class A amplifiers. The use of transistor push-pull driver and output circuits has eliminated the need for driver and output transformers, which were an essential part of valve push-pull circuits.

In a class B push-pull output stage, two transistors are biased nearly to cut-off, so that only a small quiescent current flows under zero-signal conditions. The signal is applied in antiphase to each transistor by means of some kind of phase-splitting arrangement, so that one transistor is cut off while the other conducts. With push-pull operation, the efficiency is 60 to 65% and the average current drain is low compared with class A operation. This low current drain is, of course, a great advantage in portable battery receivers.

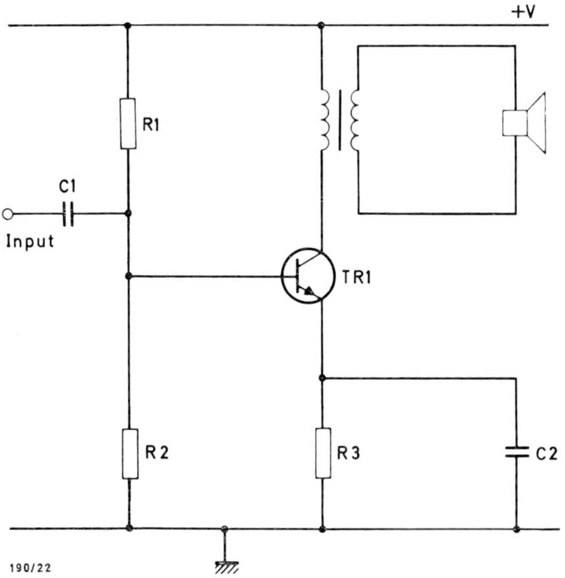

Fig. 22—Basic class A output stage

If both transistors are biased exactly to cut-off, a form of distortion occurs which is extremely unpleasant to the listener. This is called 'cross-over' distortion and is overcome by applying a slight forward bias to each transistor.

Output transistors dissipate considerable amounts of power and are, therefore, usually fitted with some kind of heatsink.

Because push-pull circuits using complementary pairs have virtually superseded those using two similar transistors in all but the output stages of high-power amplifiers, they are considered first.

Complementary Push-Pull Circuits

A basic complementary push-pull output stage is shown in Fig. 23. The signal from the driver transistor TR_2 is applied simultaneously to the bases of the two output transistors, TR_3 and TR_4, which conduct during alternate halves of the waveform. The amplified signal, occurring as variations in the potential at point A, is applied via capacitor C_4 to the loudspeaker. The collector load of the driver transistor is returned to the live end of the loudspeaker so that the signal is applied between base and emitter of the output transistors. The output devices therefore operate in common-emitter, giving maximum amplification.

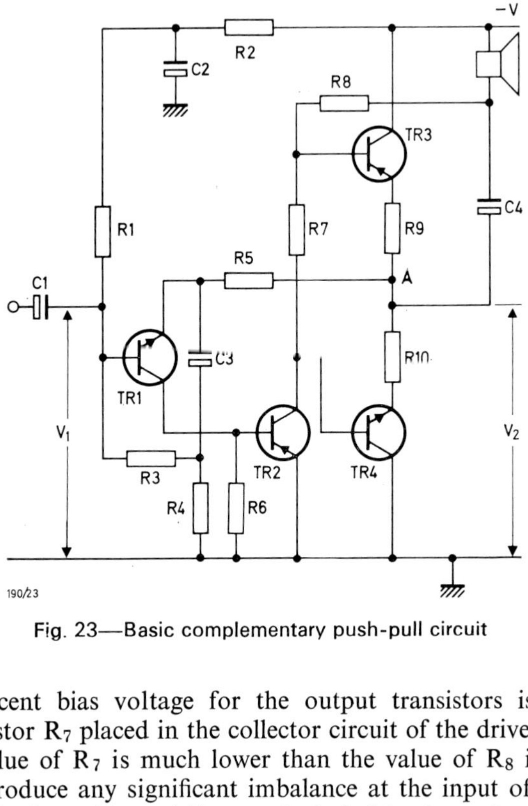

Fig. 23—Basic complementary push-pull circuit

The quiescent bias voltage for the output transistors is developed across a resistor R_7 placed in the collector circuit of the driver transistor. Since the value of R_7 is much lower than the value of R_8 its inclusion does not introduce any significant imbalance at the input of the output transistors. Resistors R_9 and R_{10} are included in the emitter circuits of the output transistors to improve thermal stability.

Transistor TR_1 acts as a pre-amplifier for a.c. signals and as a d.c. difference-amplifier comparing the voltage V_1, derived from the potential divider R_1, R_2, R_3, with voltage V_2. The high loop gain of the circuit keeps the small difference between V_1 and V_2 constant, so that V_2 is defined with respect to V_1 regardless of spreads in the characteristics of transistors TR_1 and TR_2 and the tolerance variations in values of R_5, R_6, R_7 and R_8.

Negative feedback is taken via resistor R_5 from the output to the emitter of TR_1. Decoupling is provided by capacitor C_3 to remove a.c. feedback between these two points, and a known small amount of feedback is re-introduced by means of R_4 to reduce gain spreads and distortion.

Crossover distortion can be eliminated by biasing the output transistors to give a linear combined transfer characteristic. For a given output stage, an optimum bias condition (and quiescent current) exists for which the crossover distortion is a minimum. Changes in supply voltage or ambient temperature, however, will cause departures from this optimum and therefore increase the crossover distortion.

Considerable spreads about the nominal current may result from tolerance spreads in resistor values and transistor characteristics, and from variations in ambient temperature and supply voltage. The effects of changes in ambient temperature can be minimised by the use of a thermistor in conjunction with the biasing resistor, R_7, which can be preset to compensate for component spreads. The effects of supply voltage changes can be minimised by means of a diode or other stabilising device from which the bias can be derived.

Zobel Network

The purpose and operation of the Zobel network can be explained by considering the simple complementary-pair output stage shown in Fig. 24.

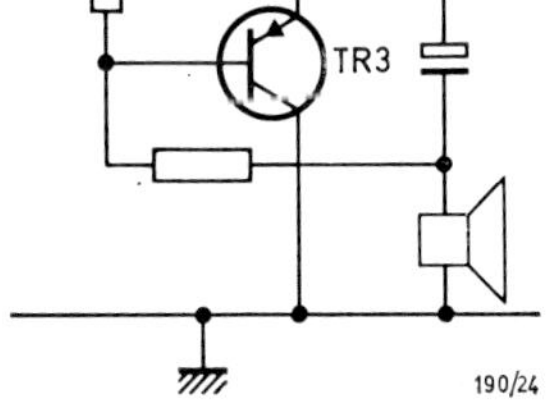

Fig. 24—Basic complementary-pair output stage

The operating limits of this circuit are defined by the continuous line in Fig. 25. While the load line remains within the enclosed area, the transistors will not be damaged by overloading. With a purely resistive load, this is a fairly easy matter to arrange. However, if the load is inductive, the limits could be exceeded during transients when the amplifier receives a large signal.

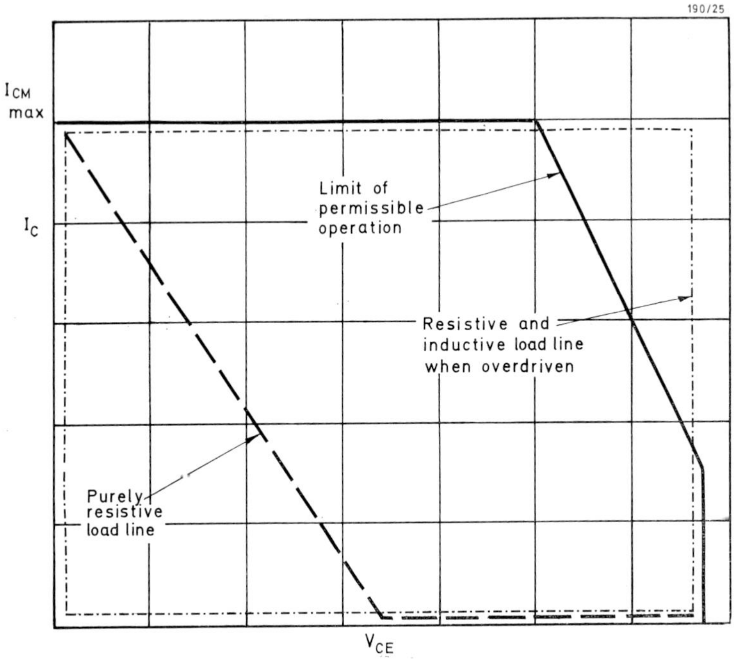

Fig. 25—Operating limits and load lines for a transistor in a complementary pair

In extreme cases, too large a signal to the driver approximates to a large squarewave. When such a signal first bottoms the driver transistor, current flows from TR_2 into the load. On the next sharp edge of the input signal, TR_1 is rapidly cut off. With a purely resistive load, TR_2 is cut off, TR_3 conducts, and the current reverses immediately. With an inductive load, however, the inductance opposes the change of current and the potential at point A rapidly decreases.

If the line voltage is less than the maximum permissible voltage of TR_2, the potential at A will become slightly negative until TR_3 acts as a reverse transistor. The energy stored in the inductance is then safely dissipated before breakdown occurs.

However, if the line voltage exceeds the maximum permissible voltage of TR_2, the voltage across TR_2 will reach the maximum permissible voltage before A becomes negative. Consequently, the energy stored in the inductance will be dissipated in the breakdown region of TR_2 and may be sufficient to cause failure of the transistor. Failure of TR_2 will be followed by failure of TR_3 as this will then be subjected to the full line voltage.

As nearly all loudspeakers contain inductance, output circuits must be designed to allow for it. This can be done either by using transistors with high breakdown voltages or by taking precautions to prevent the conditions for transistor breakdown being satisfied. If the transistors were presented with a purely resistive load, the breakdown of TR_2 would not occur. This can be achieved by connecting a Zobel network in parallel with the loudspeaker as shown in Fig. 26. The Zobel network is a capacitance, C_Z, in series with a resistance, R_Z, the values of these components having a certain relationship to the inductance and resistance of the loudspeaker.

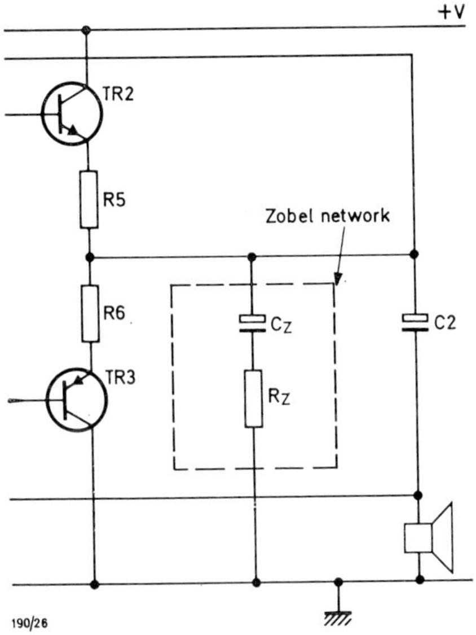

Fig. 26—Basic complementary push-pull circuit with Zobel network

If the loudspeaker is equivalent to a resistance, R_S, in series with an inductance, L_S, the components of the Zobel network must have values given by:

$$R_Z = R_S, \text{ and } C_Z = \frac{L_S}{R_S{}^2}.$$

When these conditions are satisfied, the parallel combination of the loudspeaker and the Zobel network will present a purely resistive load to the output transistors. In practice, of course, the impedance of the loudspeaker is only approximately a resistance in series with an inductance. Consequently, the use of a Zobel network will not result in a perfect resistive load, but the similarity is very close.

Push-pull Circuits using Two Similar Transistors

The complementary push-pull circuit is normally used for amplifiers with low output powers, but the push-pull circuit using two similar transistors is still often used as the output stage of high-power amplifiers. This is because n-p-n silicon transistors capable of handling high powers are more readily available than p-n-p transistors. The most usual arrangement for high-power amplifiers is for an output stage consisting of two n-p-n transistors connected in a push-pull configuration to be driven by a complementary push-pull circuit as shown in Fig. 27 (quasi-complementary configuration).

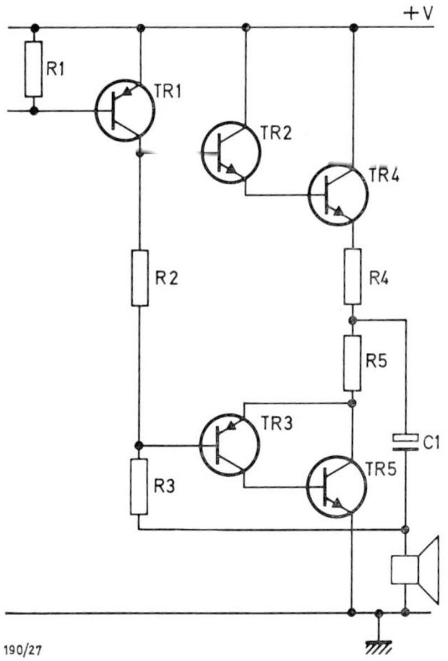

Fig. 27—Push-pull output stage using two n-p-n transistors driven by a complementary push-pull circuit

π-Mode Operation

Class AB operation is that in which the values of bias and signal voltage applied to the transistor cause collector current to flow for appreciably more than a half, but less than a whole, cycle of the signal voltage.

A class AB amplifier which operates in the π-mode differs from a normal class AB amplifier in that the total current remains constant with drive. Increasing the level of drive causes the operating point of the transistors to change first from class A to AB and then to class B.

With π-mode operation, the problem of crossover distortion associated with straight-forward class B operation does not arise. Regulation of the power supply is not important since, with π-mode operation, the current drain from the supply is constant and adequate smoothing of the power supply can be obtained by simple RC filtering. Distortion at normal listening levels is very low since the transistors operate in class A at low power levels and, with the use of emitter-loaded operation, the distortion of the amplifier over the full power range can be controlled to specific limits by a suitable choice of driver-stage source impedance.

The chief disadvantage of π-mode operation is that it requires a relatively large number of transistors to give the required power output.

CLASS D OPERATION

Class D operation differs from the modes of operation already considered in that the output transistors are used as switches in a single-ended push-pull arrangement. The transistors are fed with a squarewave which carries the audio information as pulse-width modulation. The transistors switch the load between positive and negative voltages, the duration for which it is switched to either supply depending on the mark/space ratio of the input waveform. The transistor output is coupled to the load through a low-pass filter which removes the high-frequency content and feeds an audio-frequency waveform to the load.

This arrangement makes use of the fast switching times and low saturation voltages of the modern switching transistor. It offers potential benefits over other modes of operation: only the amount of power supplied to the load at any instant is drawn from the power supply and no excess power is converted to heat by the output transistors; the linearity of the circuit is independent of the transistors; large component tolerances are acceptable because the transistors alternate between fully 'off' and fully 'on', and no biasing currents are required; efficiencies of 95% can be realised and a large range of load impedances can be connected without substantially affecting efficiency. However, the high-speed circuitry and radiation due to fast switching present problems, and this concept is still in the development stage. No class D circuits are given in this book.

OUTPUT POWER RATINGS

In practical circuits, a number of different ratings may be used to define the output power from an audio amplifier; they are:

1) Speech and music: the equivalent sinewave output that will be obtained when peaks occur in low-level passages of speech or music. (Since this energy is taken from the reservoir capacitor, the impedance of the supply does not influence this figure.)

2) Sustained music: the equivalent sinewave output that will be obtained during sustained high-level passages of music. (This may be lower than the first rating because the supply voltage may fall due to regulation.)

3) Sinewave: output power attained when the amplifier is sinewave driven.

These ratings may be quoted:

(a) at a specified distortion; the output power obtained at the specified level of distortion, often 10%;

(b) at clipping; the output power at which clipping commences, that is when the working point of the transistors reaches the knee of the output characteristic resulting in a distortion of the output waveform.

The first two ratings constitute a useful measure of the output power capability of the amplifier. The third indicates the power to be obtained when the amplifier is tested with a sinewave drive and is a lower figure than the first two.

In the I.H.F.M. (Institute of High Fidelity Manufacturers) system, the following two ratings are used:

1) Continuous power: the power an amplifier is capable of delivering for at least 30 seconds with sinewave drive.

2) Music power: the power obtainable after the application of a signal, measured during a time interval so short that the supply voltages in the amplifier have not changed from their zero-signal values.

HEATSINKS

In determining the size of heatsink required for output, and sometimes driver, transistors in audio amplifiers three factors are taken into consideration:

1) the maximum permissible junction temperature of the device;

2) the thermal conductivity of the semiconductor material and encapsulation of the device, and of the heatsink material;

3) maximum ambient temperature.

The circuits given in this book are designed for operation in ambient temperatures up to 45°C and the minimum size of heatsink, to be mounted vertically in free air, has been specified for each circuit. It should be remembered that if a different heatsink material is used the size of heatsink must be adjusted accordingly.

LOUDSPEAKER

In the practical circuits which follow this chapter, the only loudspeaker specification quoted is impedance. However, it should be pointed out that the choice of loudspeaker is important in any audio system.

The factors to be considered in the choice of a loudspeaker for a particular system are discussed in Chapter 9.

CHAPTER 4

RADIOGRAMS, RECORD PLAYERS AND PORTABLE RADIOS

The chapter is divided into three sections: a.f. stages, r.f. and i.f. stages, and a complete a.m./f.m. receiver. The first section contains audio amplifiers with output powers of 1W, 2W, 3W, and 10W; a pre-amplifier for use with the 3W and 10W amplifiers; and a 3W record-player amplifier. The second section contains r.f. and i.f. stages for a.m. and a.m./f.m. receivers, and the i.f. and detector stages of an f.m. receiver.

Component lists for the circuits are given at the end of the chapter.

A.F. STAGES

1W AUDIO AMPLIFIER

The circuit diagram for a 1W transformerless audio amplifier is shown in Fig. 28. The circuit includes a complementary output pair comprising one AC127 and one AC128 transistor, an AC128 driver transistor and an AC127 pre-amplifier. The use of high-gain output transistors allows a simple amplifier, with capacitively-coupled load, to be designed with a relatively low quiescent current. The n-p-n pre-amplifier transistor contributes to very good d.c. stability.

Circuit Description

The features of this design are the economy effected in the number of components required, and the extremely good stability of the d.c. potential V_M. The latter implies less variation of output with temperature changes compared with other forms of output stage using a capacitively-coupled load.

The load is capacitively-coupled to the emitters of TR_3 and TR_4, the bases being driven directly from the collector of TR_2. The collector load R_7 of TR_2 is returned to the live end of the loudspeaker, so that the output transistors operate in common-emitter configuration, giving maximum amplification.

A small quiescent bias voltage is required for transistors operating in class B push-pull. This is developed by the collector current of TR_2 flowing through R_9. Thermistor R_8 is used to stabilise the quiescent current against temperature variation. The quiescent current is set by adjustment of potentiometer R_9.

The transistor TR_1 serves a dual purpose. It acts as a pre-amplifier for a.c. signals and also as a d.c. difference-amplifier comparing the voltage V_1 derived from the potential divider R_1, R_2, R_3, with the voltage V_M. The high loop gain of the circuit keeps the small difference between V_M and V_1 constant, so that V_M is defined with respect to V_1, irrespective of spreads in the characteristics of the transistors TR_1 and TR_2 and the tolerance variations in the values of R_5, R_6, R_7, R_8 and R_{10}.

Negative feedback is taken via the resistor R_5 from the output stage to the emitter of TR_1. Decoupling is provided by the large-valued capacitor C_3 to remove a.c. feedback between these points, and a small known amount of a.c. feedback is re-introduced by means of R_4 to reduce gain spreads and distortion.

Performance

Output power at 10% distortion	1·2W r.m.s.
Output power at 4% distortion	0·8W r.m.s.
Sensitivity (source impedance R_S = 2kΩ):	
output 50mW, 6dB feedback	2·2mV
output 1·2W, 6dB feedback	10mV
Frequency response (—3dB points)	100Hz to 20kHz
Driver current	7·7mA
Peak collector current	470mA
Total quiescent current	13mA

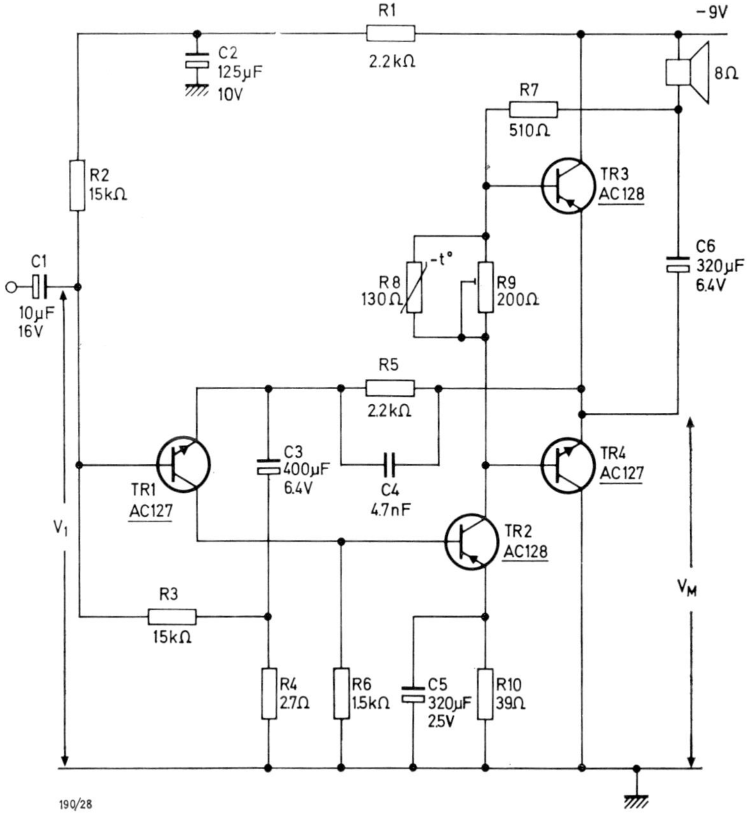

Fig. 28—1W audio amplifier circuit

Tone Control Circuit

As feedback is applied over three stages in the amplifier of Fig. 28, negative-feedback tone controls can cause instability and passive controls at the input should be used. Fig. 29 shows a simple arrangement for obtaining variable bass and treble cut, without significant loss of gain at mid-frequencies.

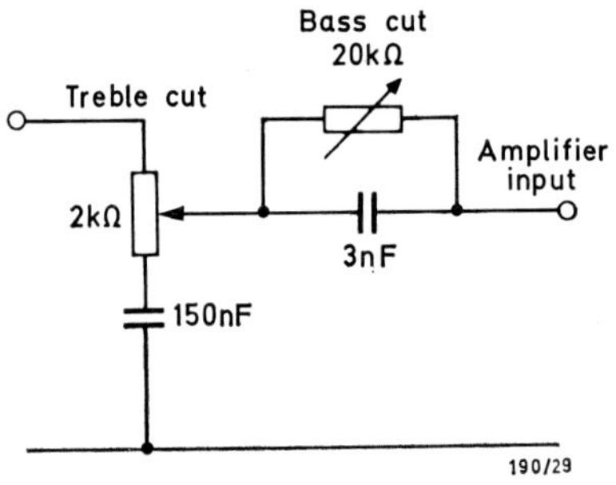

Fig. 29—Tone control circuit

2W AUDIO AMPLIFIER

Circuit Description

A 2W audio amplifier can be very simply constructed using a TCA160 integrated circuit and a few peripheral components. The circuit is shown in Fig. 30. The amplifier operates from a +12V supply to deliver 2W output power into an 8Ω speaker load at 10% total harmonic distortion. However, below 1W output power the total harmonic distortion is typically better than 1·0%. A capacitor C_1, typically 3·3nF, is connected across the input to limit the upper frequency to about 25kHz.

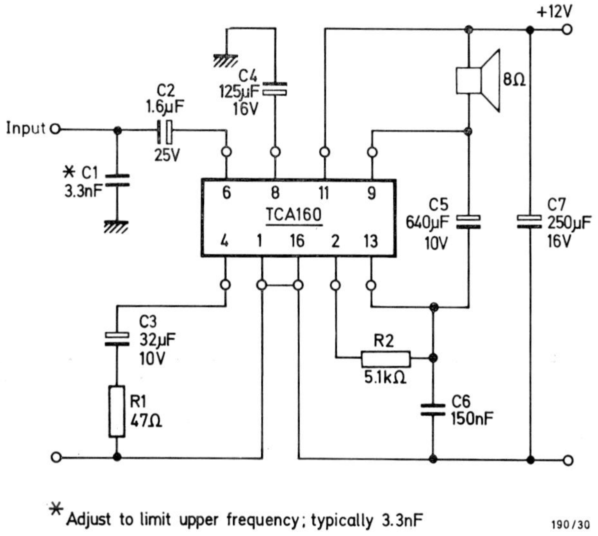

Fig. 30—2W audio amplifier circuit using TCA160 integrated circuit

Performance

Output power at 10% total harmonic distortion	2·1W
Output power at onset of clipping (typical)	1·3W
Input sensitivity (typical) for 2W output power	10mV
Total harmonic distortion (see Fig. 31):	
at 50mW output power	0·5%
at 2W output power	10%
Frequency response at onset of clipping (−3dB points) for 2kΩ source resistance	120Hz to 25kHz
Total quiescent current	5 to 15mA
Open-circuit loop gain (typical)	70dB

Signal-to-noise ratio (typical) at 2W output power with 2kΩ source resistance	75dB
Noise output power (typical)	10nW
Input impedance (typical)	15kΩ

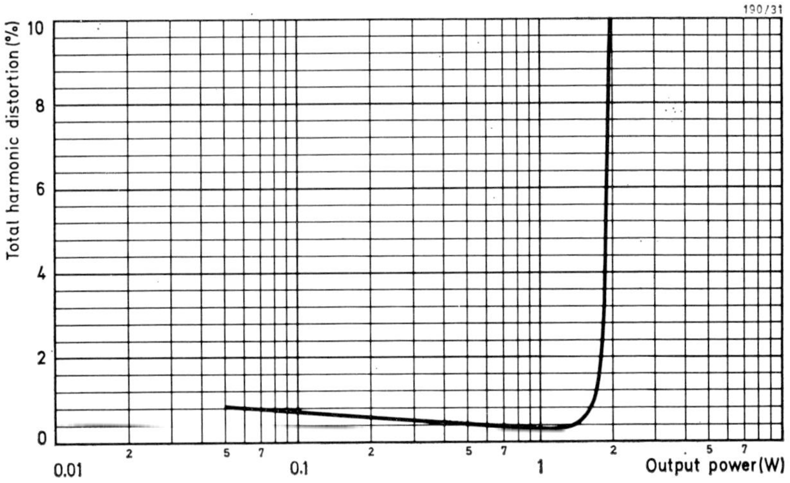

Fig. 31—Variation of total harmonic distortion with output power

Layout

To prevent an earth loop occurring between the input and output circuits, pins 1 and 16 are connected together at the IC as shown in the circuit diagram of Fig. 30. Note that capacitors C_1 and C_4 are terminated at the input earth rail and not at the output earth rail.

3W AUDIO AMPLIFIER

The circuit for a directly-coupled 3W audio amplifier is shown in Fig. 32. The complementary matched output pair comprises one AC176 and one AC128 transistor, the driver is an AC128 and the pre-amplifier is a BC148 or BC108 transistor.

Circuit Description

A conventional four-transistor configuration delivers 3W into a 12Ω load. The use of a fixed resistor R_7 for bias is acceptable because of the relatively high values of emitter resistors and the large amount of feedback. It is recommended that the resistors which determine the working point—that is resistors R_1, R_2, R_3, R_7, R_8—should be 5% tolerance components.

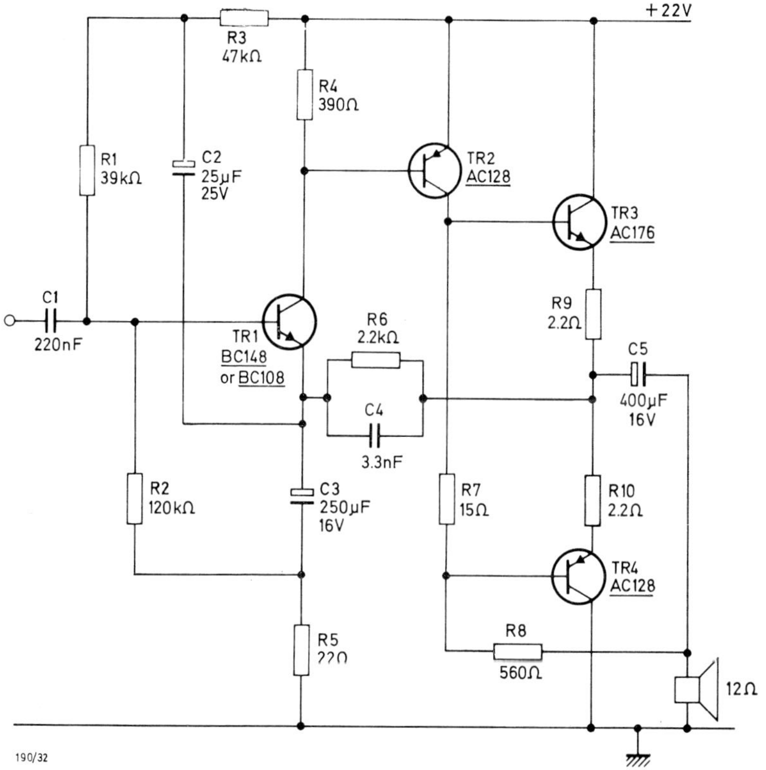

Fig. 32—3W audio amplifier circuit

Power Supply

A suitable power supply is shown in Fig. 33; the requirements are:

Quiescent consumption	25mA
Full power consumption	200mA
Effective output resistance	⩾ 20Ω

Performance

Output power	3W r.m.s.
Sensitivity for 3W output	70mV
Frequency response (—3dB points)	40Hz to 20kHz
Input resistance	80kΩ
Feedback	20dB
Distortion at 3W output (see Fig. 34)	1·7%

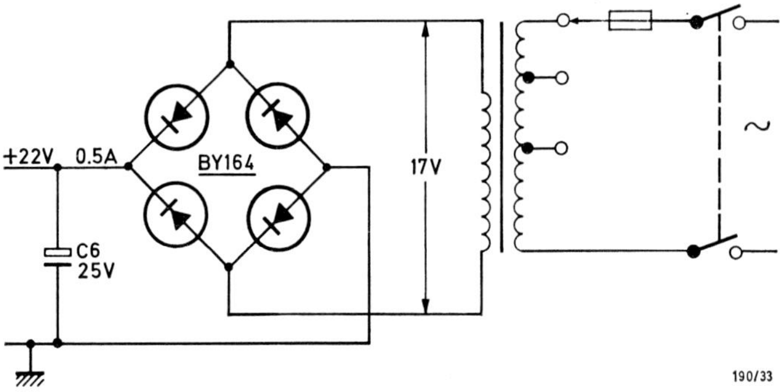

Fig. 33—Power supply circuit for 3W amplifier

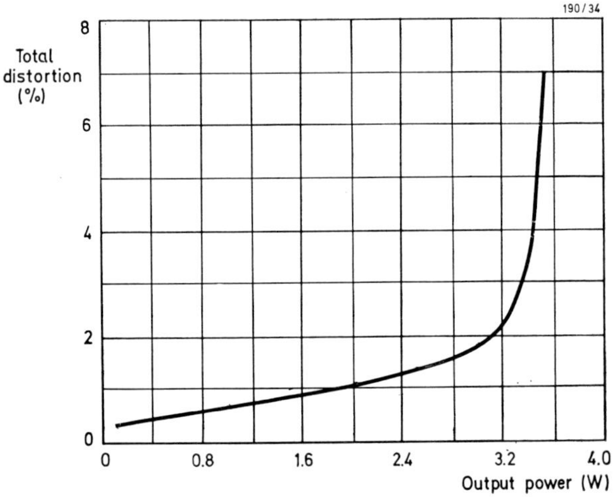

Fig. 34—Variation of total harmonic distortion with output power

10W AUDIO AMPLIFIER

The circuit of a 10W amplifier is shown in Fig. 35. In this circuit, a complementary matched pair push-pull output stage is used, and a Zobel network is incorporated.

Circuit Description

The emitter of transistor TR_1 is connected to the output stage via R_5 to provide a path for negative feedback. Bypass capacitor C_6 limits the feedback at high frequencies. The quiescent current of TR_3 and TR_4 is set to approximately 15mA by means of R_9. The thermistor R_8 compensates for temperature changes and maintains a constant quiescent current. The amplifier can deliver an output of 10W into a 4Ω loudspeaker with a distortion of 2% (see Fig. 37).

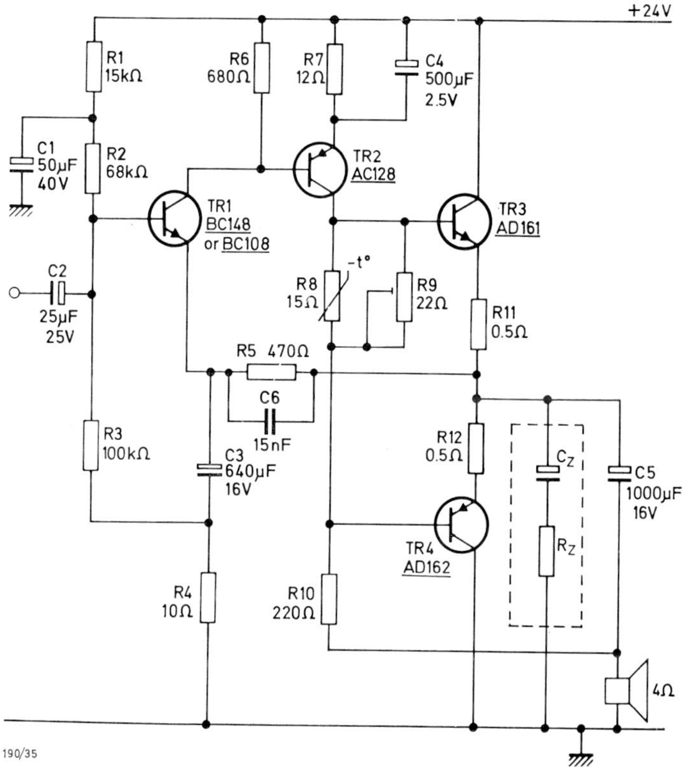

Fig. 35—10W audio amplifier circuit incorporating a Zobel network

Zobel Network

The purpose and operation of the Zobel network is explained in Chapter 3. From the formulae given on page 27, and for a 4Ω loudspeaker with an inductance of 200μH:

$$R_Z = 4\Omega,$$

$$C_Z = \frac{200 \times 10^{-6}}{16} = 12{\cdot}5\mu F.$$

(A 25V component is required.)

Power Supply

The nominal line voltage required by the amplifier is +24V and the absolute maximum voltage is +27V. The maximum value of direct current per channel is 0·8A.

The circuit of a simple power supply suitable for use with a stereo system is shown in Fig. 36. The mains transformer has a secondary output of 18V which is rectified by a diode bridge and smoothed by capacitor C_7.

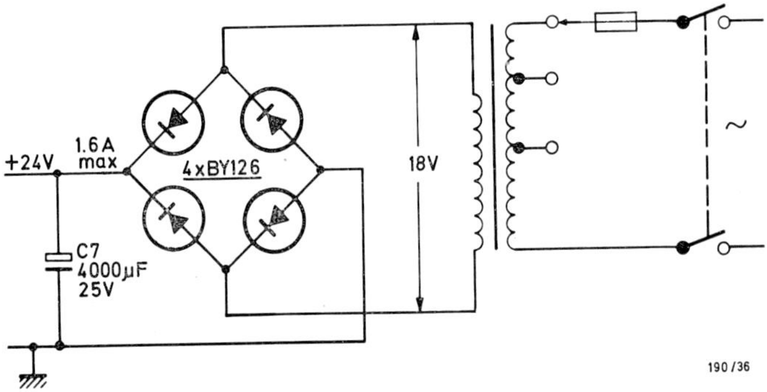

Fig. 36—Power supply circuit for 10W amplifier

Performance

Output power	10W r.m.s.
Sensitivity (voltage at base of TR_1 for rated output)	130mV
Frequency response (−3dB points)	50Hz to 16kHz
Input impedance	40kΩ
Distortion at 10W output (see Fig. 37)	2·1%

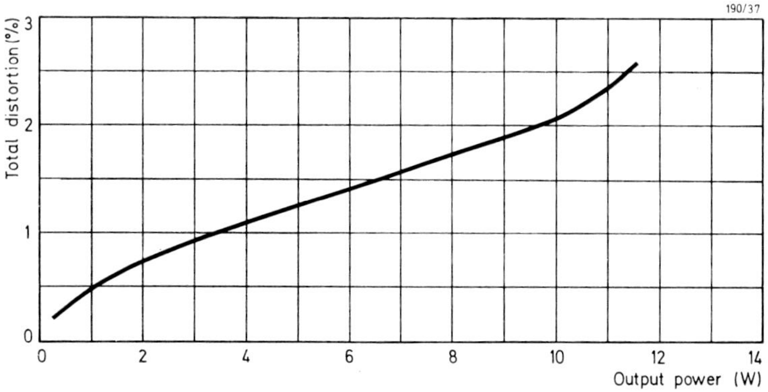

Fig. 37—Variation of total harmonic distortion with output power

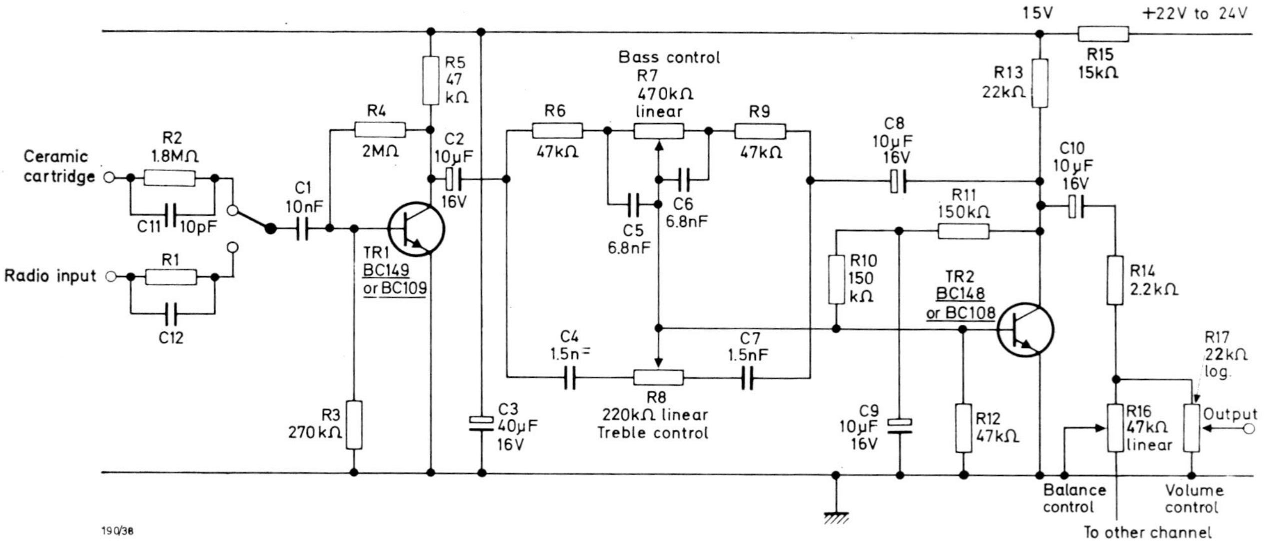

Fig. 38—Pre-amplifier circuit for use with 3W and 10W amplifiers

PRE-AMPLIFIER FOR USE WITH THE 3W AND 10W AMPLIFIERS

The circuit for a two-stage pre-amplifier which may be used with the 3W and 10W amplifiers already described is shown in Fig. 38. The first stage uses a low-noise BC149 or BC109 transistor and the second stage a BC148 or BC108 transistor.

Circuit Description

The input stage provides matching for a ceramic cartridge and for radio, the resistor R_1 being chosen to give the required sensitivity in the latter case. A capacitor C_{12} is connected in parallel with R_1, the value of capacitance being chosen to give, with R_1, a time-constant of 18μs. The second stage provides comprehensive tone control, with 10dB boost and cut at 100Hz and 10kHz. This stage is followed by a simple balance control for stereo use and a volume control.

Performance

Sensitivity:	
ceramic input, for 80mV output	180mV
radio input, with $R_1 = 470k\Omega$	45mV
Signal handling, output at clipping	1·5V

3W RECORD-PLAYER AMPLIFIER

The amplifier shown in Fig. 39 is suitable for use with cartridges giving an output of 300mV at 1cm/s, and may be adapted to include comprehensive tone control (see Fig. 42). The circuit uses a complementary output pair of one AC176 and one AC128 transistor, a BC107 driver transistor and a BC108 transistor.

Circuit Description

A crystal pick-up requires a high load, 1 to 2MΩ, if the bass response is to be adequate. A simple way to provide this load, and at the same time minimise gain spreads, is to apply feedback in series with the input. This technique is used in this amplifier.

Transistor TR_1 is an emitter follower supplying the driver transistor

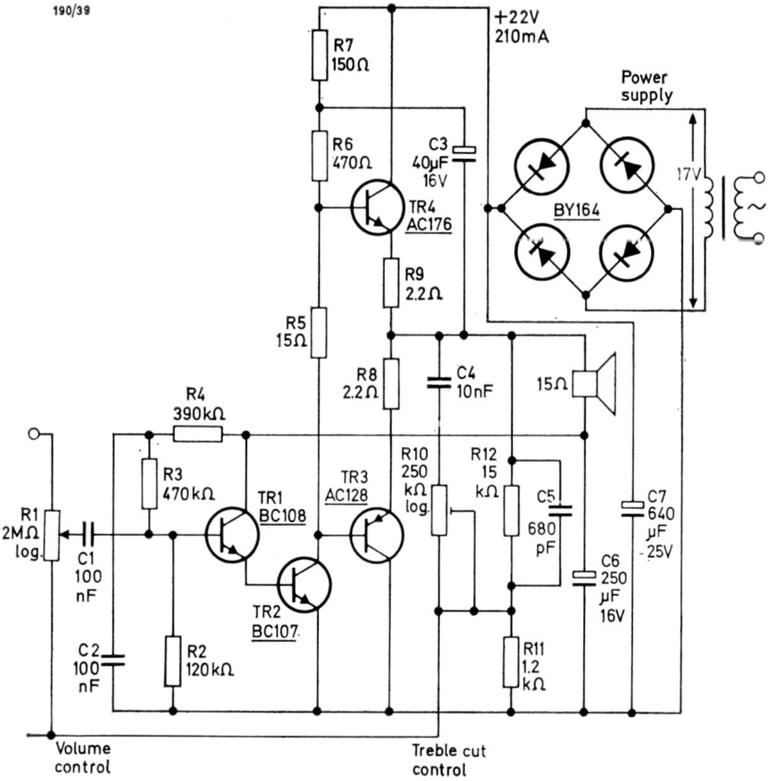

Fig. 39—3W record player amplifier circuit

TR_2. Because of the large amount of feedback used in the circuit, the output transistors, TR_3 and TR_4, can have simple resistive bias. The output voltage is divided by resistors R_{11} and R_{12} and fed back to the input. The network C_4 and R_{10} forms a simple treble cut control.

The loudspeaker and capacitor C_6 are interchanged with respect to their conventional positions to provide a low-voltage supply for the collector of transistor TR_1, and also a point relatively free from a.c. signals from which d.c. feedback can be taken. The output mid-point voltage is compared with the base voltage of transistor TR_1 by the potential divider comprising resistors R_2, R_3 and R_4. Although the voltage is taken from a nominally d.c. point, decoupling is still required to remove low frequencies.

Performance

Output power	3W r.m.s.
Sensitivity for full output	600mV
Input resistance:	
with volume control at maximum	660kΩ
with volume control at minimum	2MΩ
Feedback, depending on position of volume control, up to	25dB
Distortion at 2W output:	
with volume control at maximum	0·4%
with volume control at −10dB	2%
Quiescent current	20mA
Total current at 3W output	210mA

Fig. 40 shows the variation in bass response with volume control setting, due to the variation in input resistance. It should be noted that the circuit given in Fig. 40 representing the pick-up was used to obtain these curves.

Fig. 41 shows the variation of distortion with output power. With the volume control at maximum setting the feedback is high and the distortion low. However, at intermediate settings of the volume control the feedback is reduced, because of the increase in source impedance, and the distortion is higher.

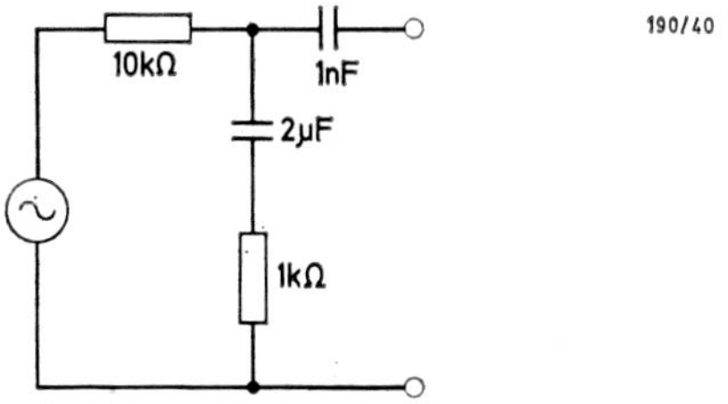

Cartridge equivalent circuit

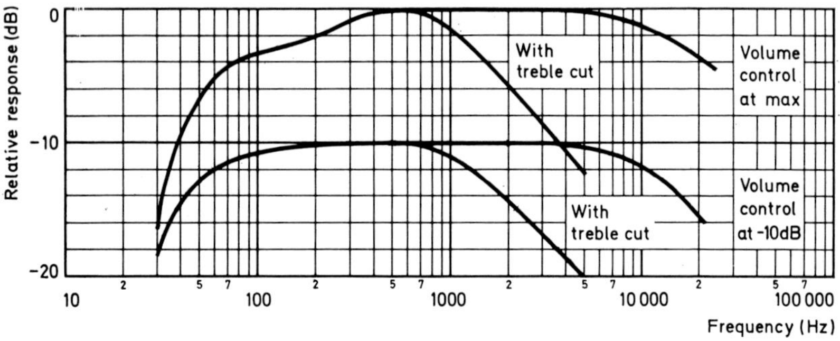

Fig. 40—Frequency response of amplifier

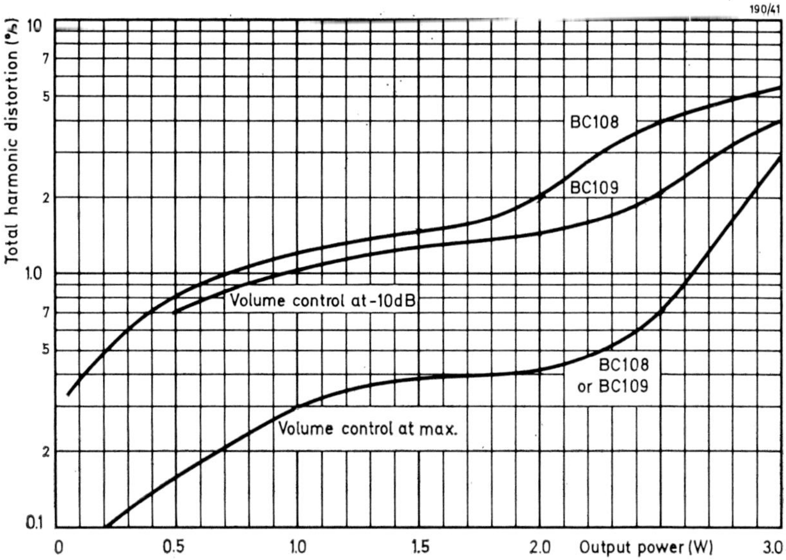

Fig. 41—Variation of total harmonic distortion with output power for basic amplifier and modified circuits

Circuit adapted to include Comprehensive Tone Control

The feedback circuit may be modified to give bass and treble cut and boost controls, as shown in Fig. 42.

The input transistor TR_1 is a BC109, since this device gives a useful improvement in bass response and also in distortion (see Fig. 41). The tone control characteristics with volume control at maximum and at −10dB are shown in Figs. 43 and 44. The value of capacitor C_1 is increased to ensure low-frequency stability with bass boost applied. Capacitor C_9 is included to block d.c. from the bass control, and so prevent noisy operation.

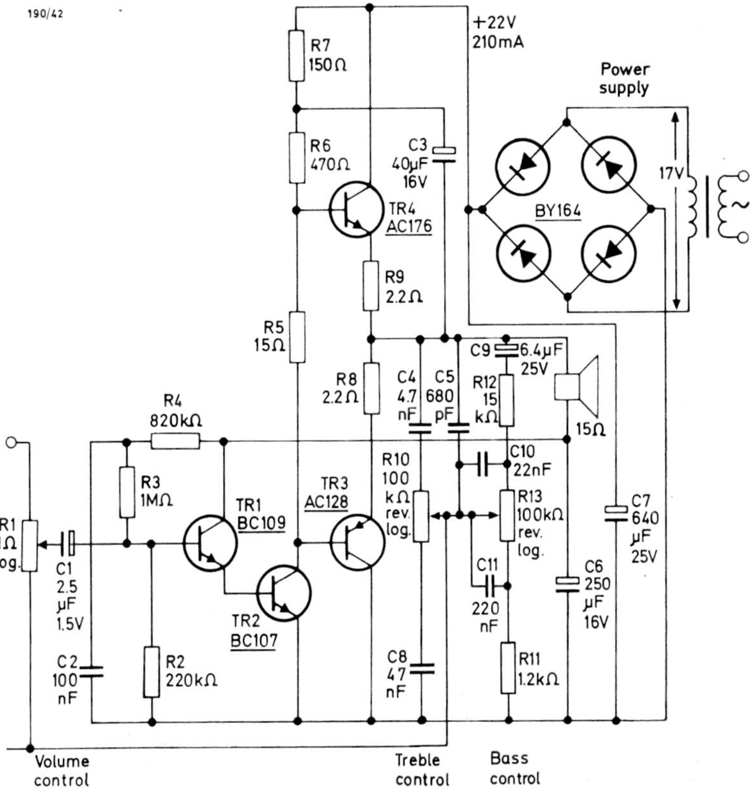

Fig. 42—Amplifier circuit modified to include tone control

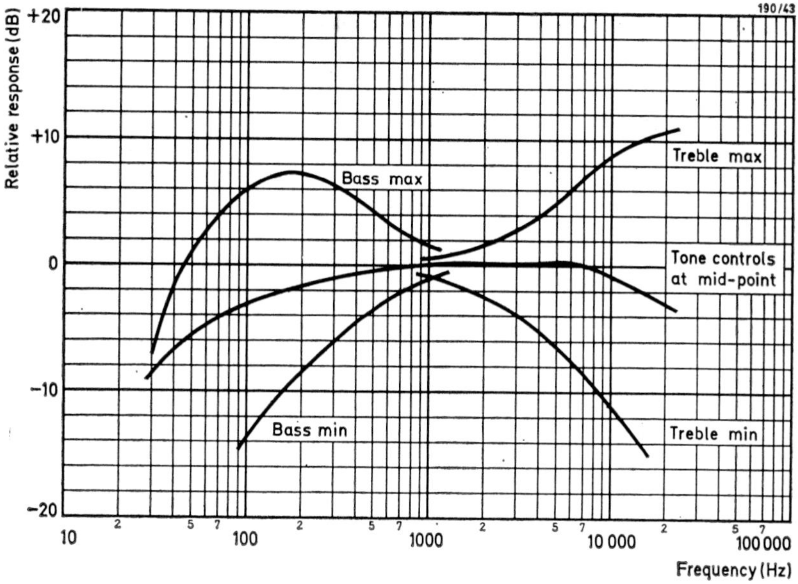

Fig. 43—Tone control characteristics with volume control at maximum

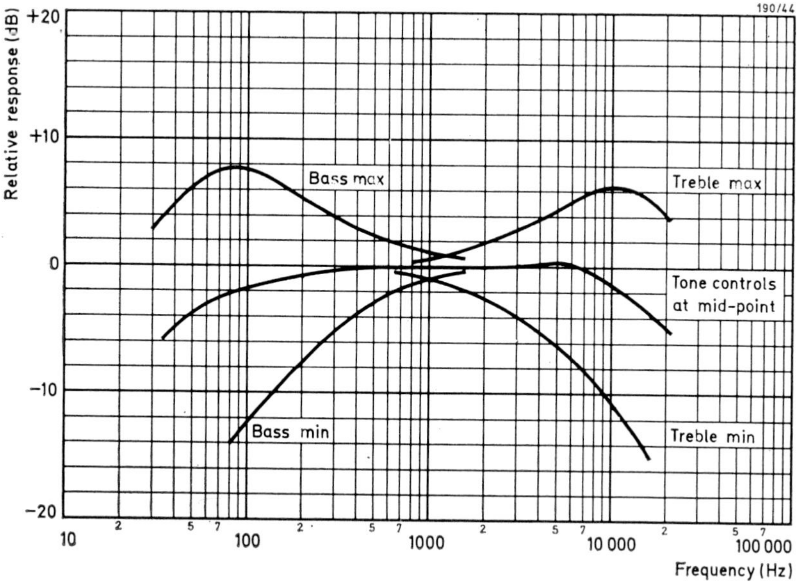

Fig. 44—Tone control characteristics with volume control at −10dB

R.F. AND I.F. STAGES

Attention is drawn to the difficulties associated with the construction and alignment of r.f. and i.f. stages. As stated earlier, access to specialised equipment is essential to those wishing to build these circuits.

A.M. RECEIVER

The r.f. and i.f. stages of a nine-volt medium-wave receiver are shown in Fig. 45. The circuit uses Lockfit transistors, types BF194 and BF195*.

Circuit Description

Self-oscillating Mixer Stage

The self-oscillating mixer is a conventional circuit, and the BF194 transistor is operated with a collector-emitter voltage of 4·8V, and an emitter current of 1mA. Voltage biasing is used to stabilise the operating point against h_{FE} and V_{BE} spreads; the emitter RC circuit is also necessary to stabilise the oscillator voltage.

The aerial source impedance is 400Ω, and the i.f. load on the mixer is 9kΩ. The working Q-factor of the first i.f. transformer is 80.

First I.F. Stage

The first i.f. transistor is a BF195, and is current-biased from a tap on the detector-diode forward-bias chain. It operates at a collector-emitter voltage of 7·5V, and an emitter current of 1mA.

To ensure good large-signal handling properties, a very large step-down ratio is used between the mixer load and the first i.f. stage. (The signal-handling ability is limited by the amplitude of the signal which can be accommodated at the base of the gain-controlled stage.) Because of the large step-down ratio, the source impedance of the first i.f. stage is very low (21·5Ω). Since this stage drives a load of 50kΩ, it has a very high stability factor (24). However, the gain is sufficiently high, and any increase in load impedance would result in gain spreads caused by variations in output conductance.

Second I.F. Stage

The second i.f. transistor is a BF195, and is current-biased. It operates at a collector-emitter voltage of 7·5V, and an emitter current of 1mA.

The load impedance of the second i.f. stage is 7·5kΩ, and this value ensures maximum power for the a.g.c. line. The source impedance is 530Ω, and therefore the stability factor of the second i.f. stage is 6·5.

---

*The BF194 and BF195 are supplied to receiver manufacturers in sets of three devices. The devices in these sets carry the letters B, C or D added to the normal type number. For replacement or home-construction purposes, these letters should be ignored and the required devices ordered simply by their type numbers, BF194 and BF195.

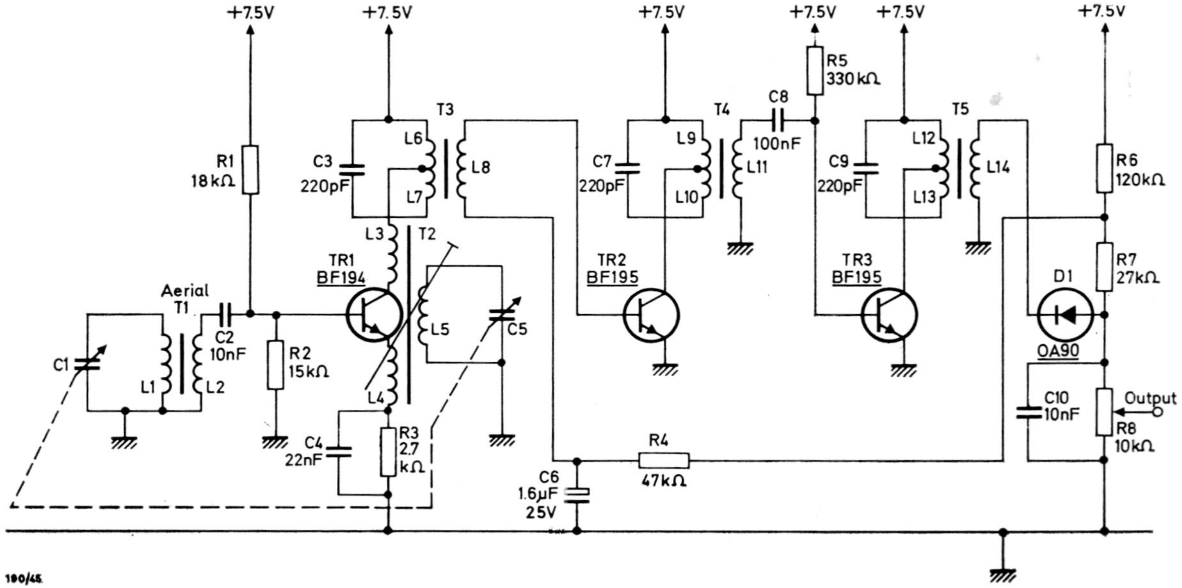

Fig. 45—R.F. and i.f. stages of an a.m. receiver
(Line-up using sets of transistors: TR_1—BF194B, TR_2—BF195C, TR_3—BF195D)

Detector

The OA90 germanium detector diode is forward-biased by 15μA to improve the small-signal efficiency, and to ensure constant loading on the last i.f. transformer.

Performance

A.G.C. range for 6dB change in output (see Fig. 46)	45dB
I.F. bandwidth (−6dB points)	±3·8kHz
Attenuation at 9kHz off-tune	28dB
Gain reduction when the supply voltage is reduced to half its nominal value	15dB
Gain spreads (95% limits) resulting from spreads in h_{FE}, V_{BE}, and 10% resistor tolerances (but excluding coil spreads)	±4dB

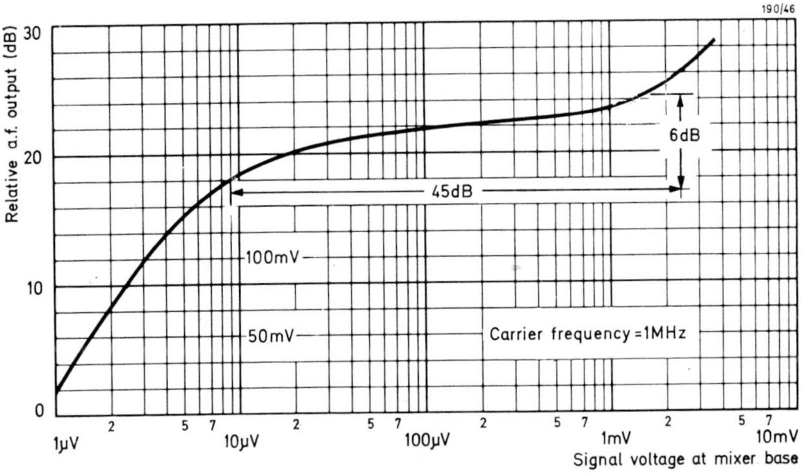

Fig. 46—Automatic gain control characteristic

A.M./F.M. RECEIVER

The r.f. and i.f. stages of an a.m./f.m. receiver operating from a 9V supply are shown in Fig. 48 (pages 56 and 57). The design uses three BF195 and two BF194 transistors.

Circuit Description

F.M. Tuner

The f.m. tuner is conventional in design, using a BF195 operated at 1mA as an r.f. amplifier. The self-oscillating mixer uses a BF195 operated at 2mA, the best compromise to obtain minimum oscillator drift with any change in supply voltage (and also good operation at a reduced supply voltage).

A.M. Mixer and A.M./F.M. I.F. Amplifier

The a.m. self-oscillating mixer, TR_3, is voltage-biased using a 3·9kΩ emitter resistor in order to obtain stable operation with 100mV drive into the emitter.

The bias arrangement for the a.g.c. stage TR_4 uses d.c. feedback from the collector in order to achieve good gain stability with source voltage variations. However, current biasing of the controlled stage requires some compromise between gain stability, gain spread and a.g.c. voltage; diode D_1 is forward-biased by only 13μA to achieve good a.g.c. properties. Detector efficiency and distortion are not adversely affected by this low forward-bias current.

The third stage is voltage-biased, to assist in the achievement of a large output voltage swing at the collector.

The f.m. i.f. amplifier is of conventional double-tuned design; the a.m. i.f. amplifier uses single-tuned circuits.

A.M. Performance

6dB bandwidth	±3kHz
Selectivity at 9kHz off-tune	22dB*
Maximum signal handling	110mV/metre
Input change for 6dB change in output	46dB
Oscillator drive at 1MHz	100mV
Lower limit of supply voltage for oscillator operation	2·7V
A.G.C. and signal-to-noise characteristic	see Fig. 47

*This figure may be increased by increasing the working Q-factor at the expense of the 6dB bandwidth.

F.M. Performance

3dB bandwidth	±160kHz
Selectivity at 300kHz off tune	43dB
Image rejection ratio	26dB
Limiting sensitivity (3dB below maximum output)	22μV (aerial e.m.f.)
Maximum signal handling	20mV (aerial e.m.f.)
Oscillator drive	200mV
Oscillator drift for 2V change in supply voltage	<25kHz
Oscillator pulling with signal input, at maximum signal	<4kHz
Limiting and signal-to-noise characteristic	see Fig. 49
Total distortion	see Fig. 50
A.M. rejection	see Fig. 51

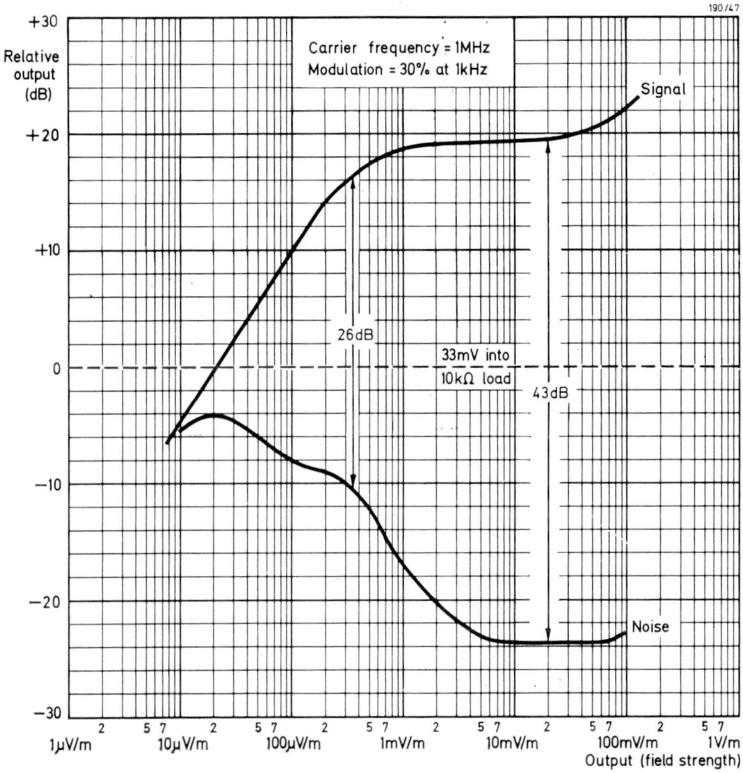

Fig. 47—Automatic gain control and signal-to-noise characteristic (a.m.)

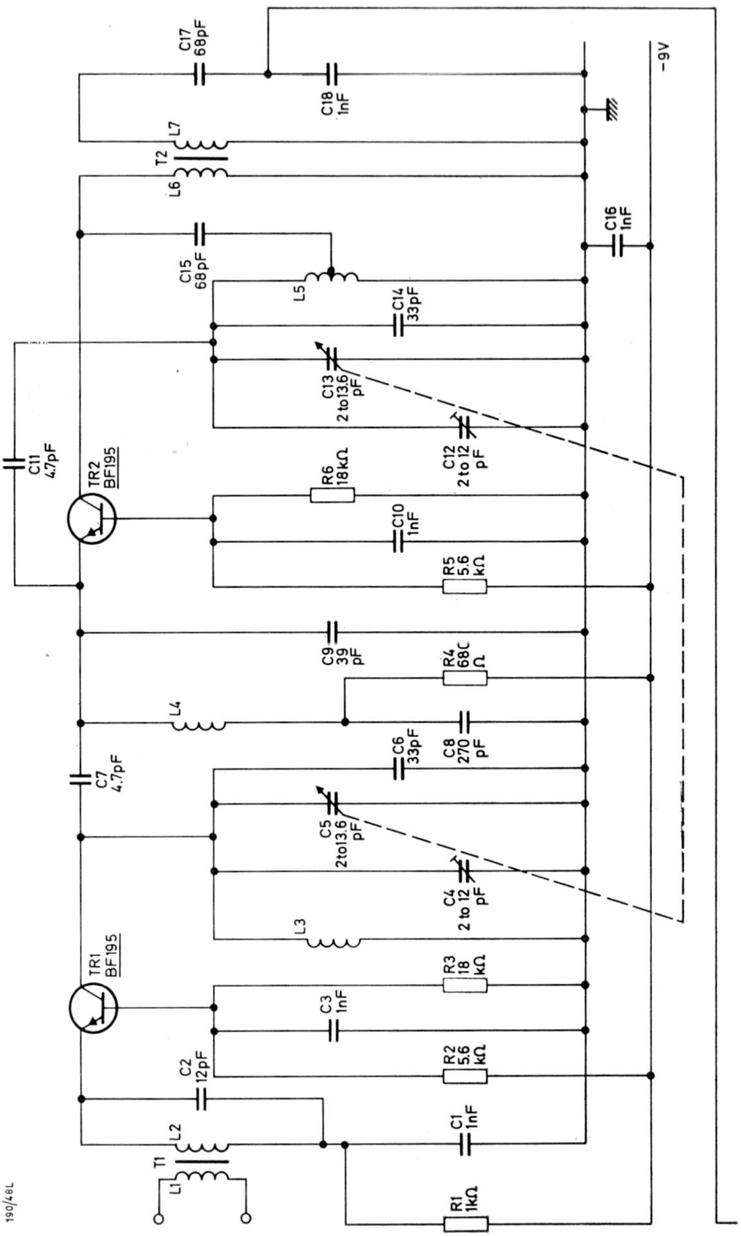

C17 68pF
C18 1nF
-9V
L7
T2
L6
C16 1nF
C15 68pF
L5
C14 33pF
C13 2 to 13.6 pF
C12 2 to 12 pF
C11 4.7pF
TR2 BF195
R6 18kΩ
C10 1nF
R5 5.6 kΩ
C9 39 pF
R4 680 Ω
L4
C8 270 pF
C6 33pF
C7 4.7pF
C5 2to13.6 pF
C4 2 to 12 pF
L3
TR1 BF195
R3 18 kΩ
C3 1nF
R2 5.6 kΩ
C2 12pF
L2
T1
L1
C1 1nF
R1 1kΩ
190/48L

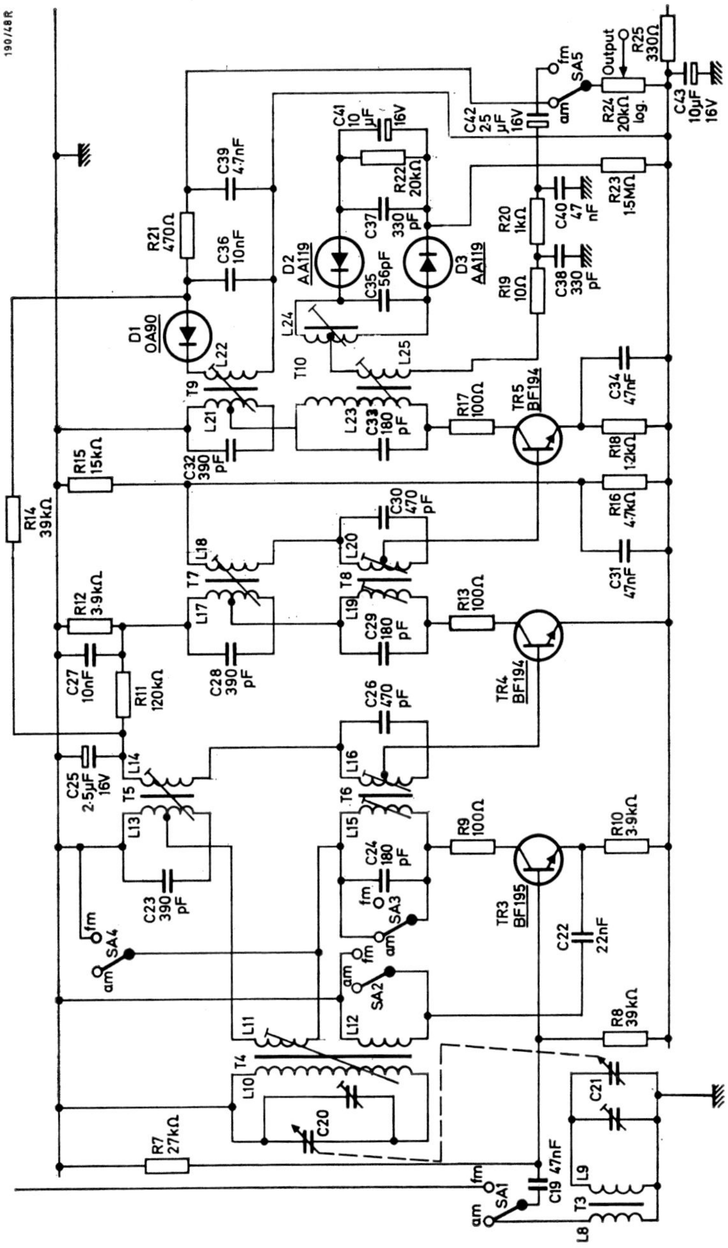

Fig. 48—F.M. tuner, r.f. and i.f. stages of an a.m./f.m. receiver

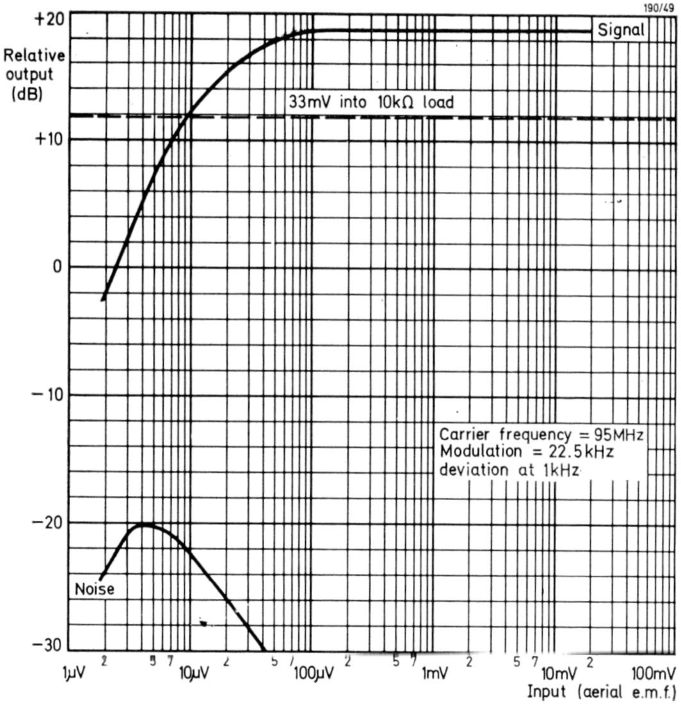

Fig. 49—Limiting and signal-to-noise characteristic (f.m.)

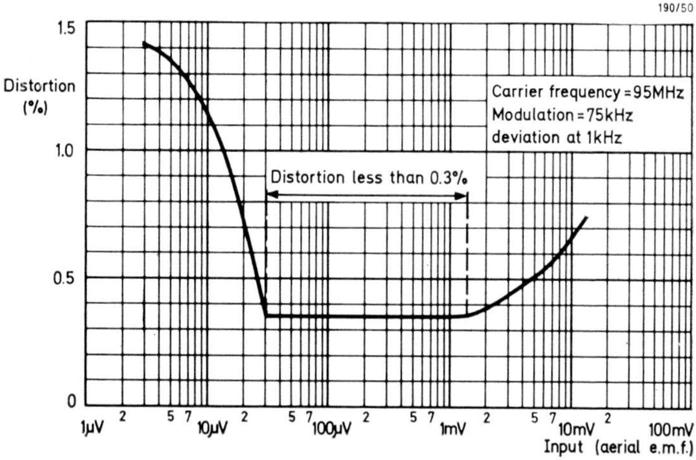

Fig. 50—Variation of audio distortion with input voltage (f.m.)

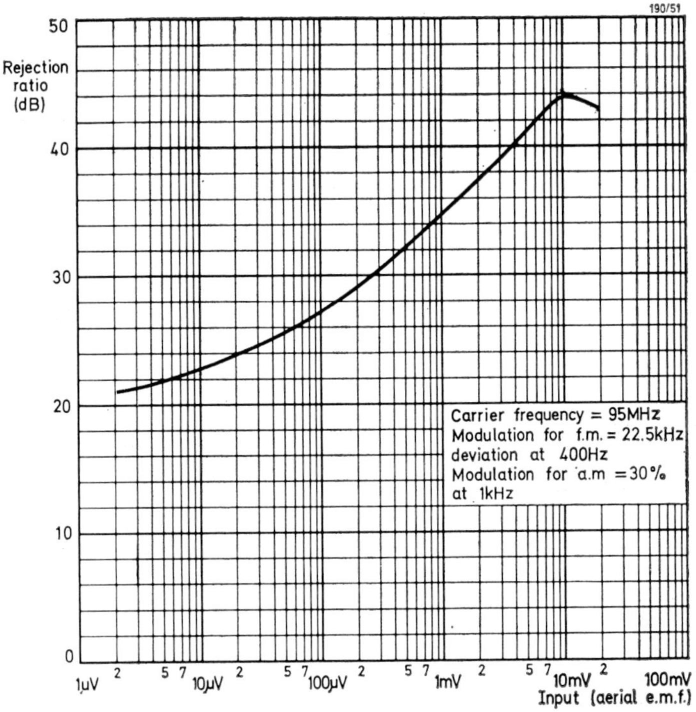

Fig. 51—A.M. rejection ratio

Transistor and Diode Currents

The quiescent collector currents of the transistors are as follows:

TR_1, TR_3, TR_4, TR_5	1mA
TR_2	2mA

The diode currents are:

D_1	13μA
D_2, D_3	5·3μA

F.M. I.F. AND DETECTOR STAGES USING INTEGRATED CIRCUIT

The circuit diagram of the 'front end' of an f.m. receiver using an integrated circuit is shown in Fig. 52. It shows clearly the simplification possible in a design by the use of an integrated circuit (IC).

The IC used, type TBA750, forms the complete i.f. amplifier and detector stages of the receiver. It also provides a volume control stage and the driver for a squelch amplifier to reduce interstation noise during tuning. Because not all the components in the IC are used, it is shown in block form in the circuit diagram. The tuner is also shown in block form as any conventional design can be used.

Circuit Description

The output of the tuner is passed through transistor TR_1 (BF194) to the 10·7MHz bandpass filter L_1 and L_2. From the filter the signal is passed to the i.f. amplifier in the IC through pins 3 and 4. The i.f. amplifier is a differential limiter type (without transformers), and provides two outputs. One is taken direct to the detector (inside the IC), while the other is taken to the phase-shifting network L_3 and C_{12} connected to pins 8 and 9 of the IC.

The detector is a balanced product type (also called a full-wave gated coincidence detector or a quadrature detector). The two inputs to the detector are 90° out of phase (the direct output from the i.f. amplifier, and that from L_3 and C_{12}), and they produce current pulses in the output whose width is proportional to the deviation in the relative phase shift between the two inputs. Thus the frequency modulation of the i.f. signal is converted into pulse-width modulation, and the a.f. signal is derived by the mean current output. The a.f. signal passes through the volume control stage in the IC, and the output signal is taken from pin 12.

The audio pre-amplifier stage in the IC is used in this design to drive a squelch amplifier which reduces the interstation noise during tuning. An a.f. signal is taken from pin 11 of the IC and fed back for amplification through C_9 and pin 1. The amplified signal is taken from pin 15, and passed to the d.c. amplifier TR_2. The output from this transistor establishes the d.c. level on pin 13 which controls the gain of the volume control stage, and on the base of transistor TR_1. The squelch amplifier is brought into operation by the off/on switch. The 330pF capacitor C_{10} connected to pins 10 and 11 provides the necessary de-emphasis.

Only one power supply of 12V is required as the supply for transistor TR_1 is derived from the IC.

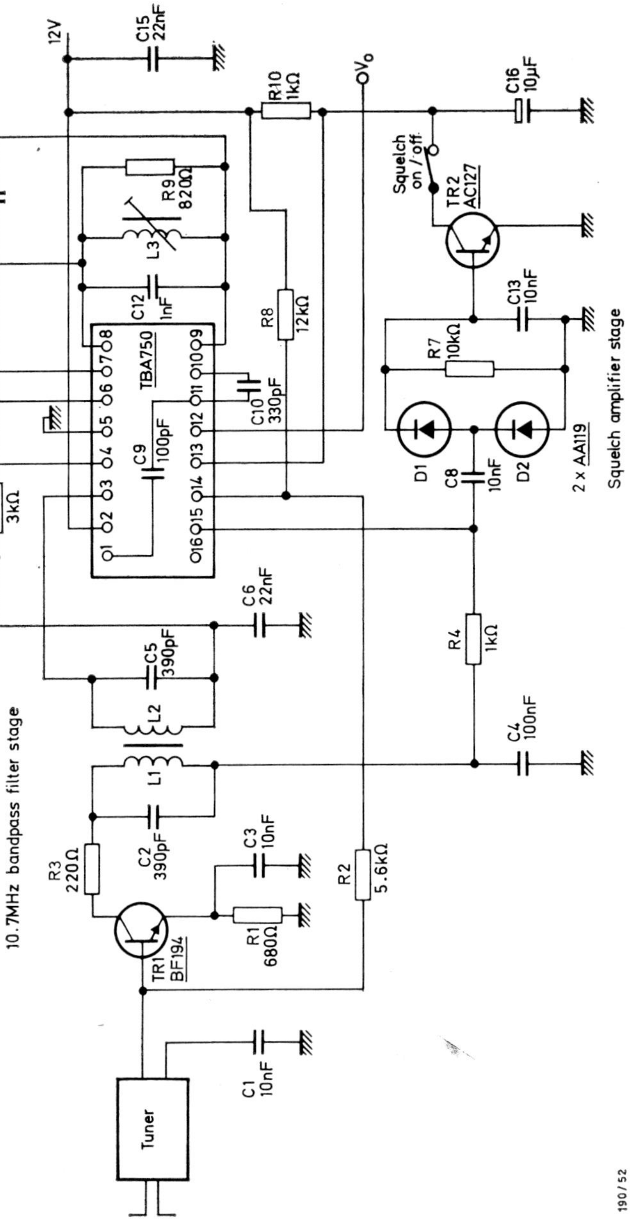

Fig. 52—F.M. i.f. and detector stages using TBA750 integrated circuit

Performance

Maximum i.f. input voltage	50mV
Signal-to-noise characteristic	see Fig. 53
A.M. rejection	see Fig. 53
Distortion, up to 50kHz deviation	1%
Control range of d.c. volume control	80dB

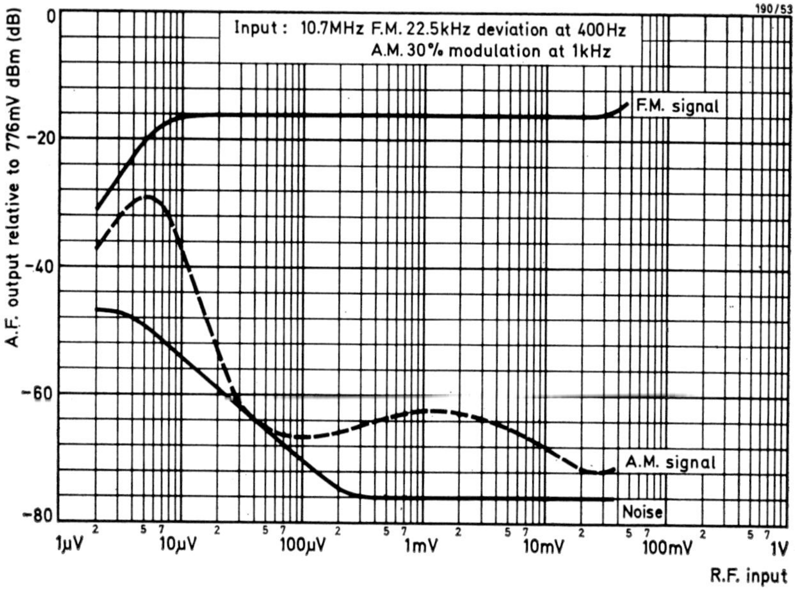

Fig. 53—Signal, noise, and a.m. rejection characteristics

AM/FM RECEIVER USING INTEGRATED CIRCUIT

The circuit diagram of a complete a.m./f.m. receiver for long, medium, and short-wave reception is shown in Fig. 54. The design uses an integrated circuit, type TBA700, in the i.f. and a.f. stages. The receiver operates from a 9V supply.

Circuit Description

The TBA700 integrated circuit provides the i.f. amplifier (apart from the coils) and the complete a.f. amplifier stage. Thus the additional sections required are the a.m. and f.m. tuners (both conventional designs), the i.f. transformers, and the a.m. and f.m. detector stages. The components contained in the IC are shown within the broken line in the circuit diagram.

Push-button wavechange switches are used in the receiver, and these are shown for f.m. reception in the circuit diagram.

A.M. Operation

Coils L_1, L_2, and L_3, wound on the ferrite rod aerial, are the tuning coils for short, medium, and long-wave reception respectively. The r.f. signal is fed to the base of the self-oscillating mixer, transistor TR_{26}. Two oscillator coils, L_6 for short-wave and L_7 for long and medium wave reception, are switched into the oscillator circuit as required by depression of the appropriate waveband switch.

The 470kHz i.f. signal is taken from the collector of TR_{26} to the first i.f. transformer formed by L_9 and L_{10}. Transistor TR_1 in the IC provides i.f. amplification. Likewise, transistor TR_2 is coupled to the second i.f. transformer L_{12} to provide i.f. amplification. The output from the final i.f. transformer L_{14} is detected by diode D_{20}.

The gain of the two i.f. amplifiers TR_1 and TR_2 is controlled by the a.g.c. signal. This is applied through pin 4 of the IC to the base of transistor TR_3 which controls TR_1 and TR_2.

The detected a.f. signal from didoe D_{20} is passed to the volume control R_{28} and on to the a.f. amplifier stages which are common to both a.m. and f.m. reception. An a.g.c. signal is developed across capacitor C_{45} which is applied to the i.f. amplifier through pin 4 of the IC.

F.M. Operation

The f.m. tuner is shown in block form as any conventional design can be used.

The output from the tuner, at 10·7MHz, is passed to the base of transistor TR_{26} which acts as an i.f. amplifier. The i.f. stage uses transformers

Fig. 54—A.M./F.M. receiver using TBA700 integrated circuit

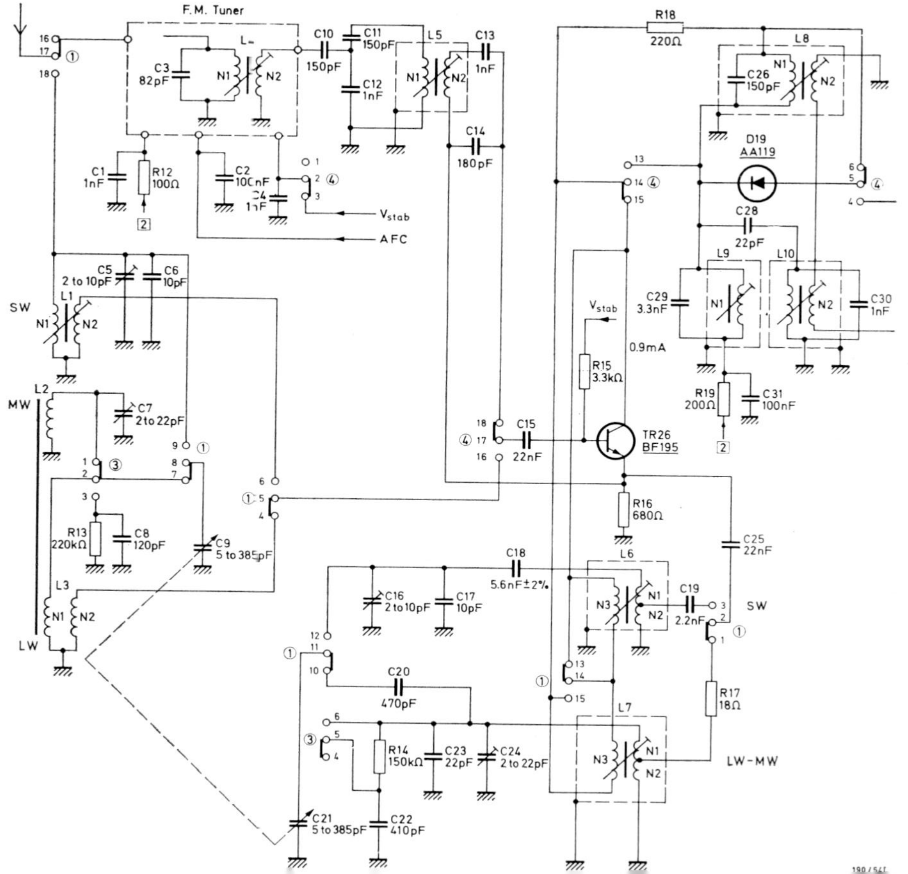

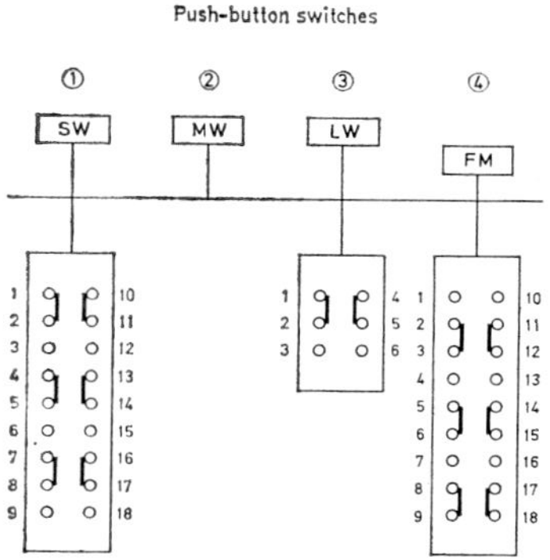

Switches in position F.M.

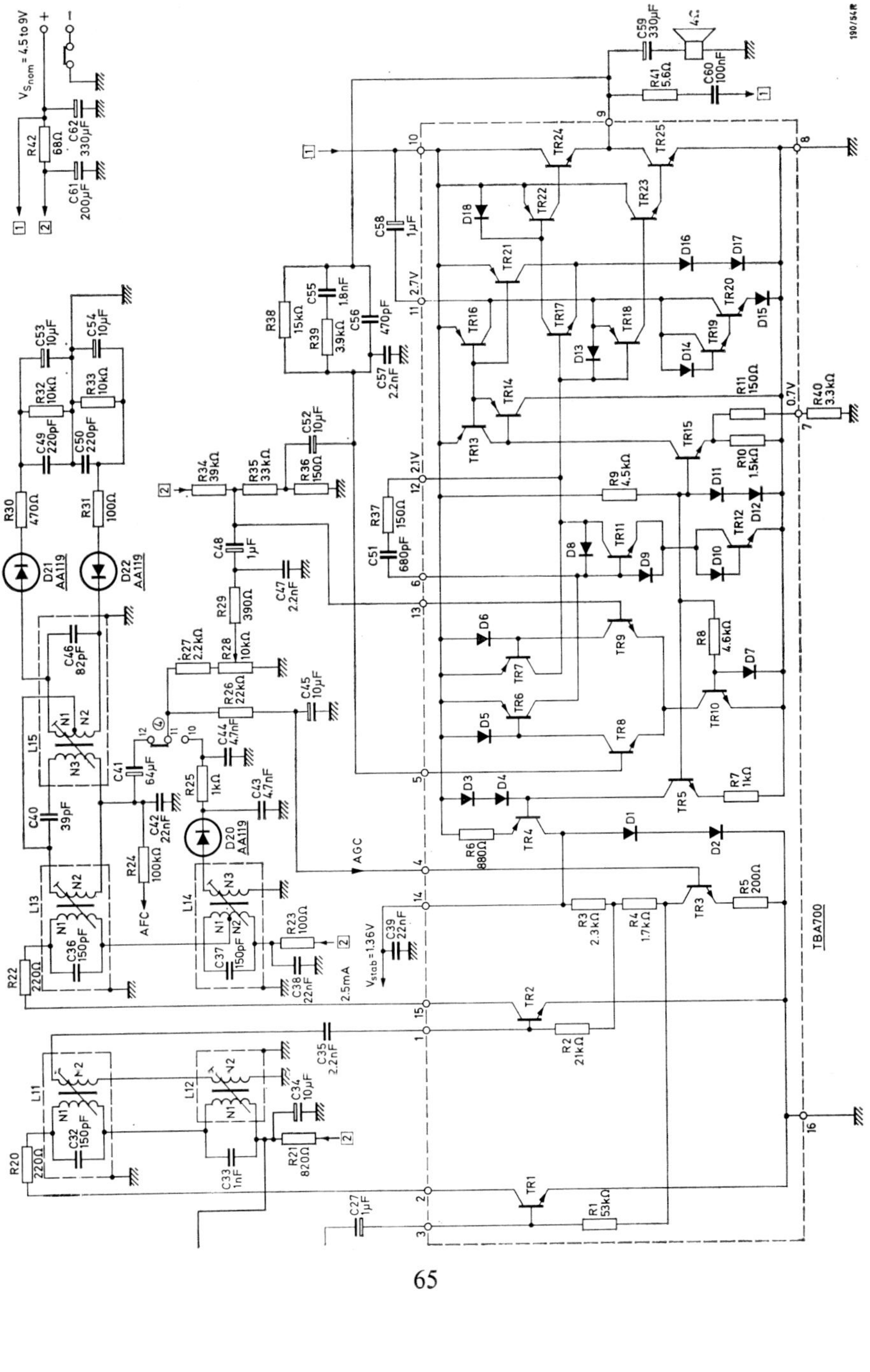

$V_{S_{nom}}$ = 4.5 to 9V
R42 68Ω
C62 330µF
C61 200µF
C59 330µF
4Ω
C60 100nF
R41 5.6Ω
190/54R
TR24
TR25
TR22
TR23
D18
C58 1µF
TR21
D16
D17
R38 15kΩ
C55 1.8nF
R39 3.9kΩ
C56 470pF
C57 2.2nF
2.7V
TR16
TR17
TR18
TR20
TR19
D13
D14
D15
TR14
R11 150Ω
0.7V
R40 3.3kΩ
TR13
TR15
R10 1.5kΩ
C53 10µF
C54 10µF
R32 10kΩ
R33 10kΩ
C49 220pF
C50 220pF
C52 10µF
R30 470Ω
R31 100Ω
R34 39kΩ
R35 33kΩ
R36 150Ω
2.1V
R9 4.5kΩ
D11
D12
R37 150Ω
C51 680pF
TR11
TR12
D8
D9
D10
D21 AA119
D22 AA119
C48 1µF
C47 2.2nF
R29 390Ω
D6
TR9
TR7
R8 4.6kΩ
D7
C46 82pF
R27 2.2kΩ
R28 10kΩ
R26 22kΩ
C45 10µF
TR6
TR10
D5
TR8
L15
N1
N2
N3
C44 4.7nF
C41 64µF
R25 1kΩ
C43 4.7nF
D3
D4
TR5
R7 1kΩ
C40 39pF
C42 22nF
D20 AA119
R6 880Ω
TR4
D1
D2
AGC
R24 100kΩ
AFC
L13
L14
C36 150pF
C37 150pF
R23 100Ω
C38 22nF
V_{stab}=1.36V
C39 22nF
2.5mA
R3 2.3kΩ
R4 1.7kΩ
TR3
R5 200Ω
TBA700
R22 220Ω
TR2
R2 21kΩ
L11
L12
C32 150pF
C35 2.2nF
C34 10µF
C33 1nF
R21 820Ω
R20 220Ω
TR1
R1 53kΩ
C27 1µF

L_8, L_{11}, and L_{13}, and i.f. amplification is also provided by transistors TR_1 and TR_2 in the IC.

The output from L_{13} is passed to the ratio detector, and the detected a.f. signal is passed to the volume control R_{28}. An a.f.c. signal is also developed from the a.f. signal across capacitor C_{42}, and is applied back to the f.m. tuner section.

A.F. Amplifier

The a.f. signal from the volume control is passed to pin 13 of the IC. It is fed to the base of transistor TR_9 which forms a long-tail pair with TR_8. The other input, to the base of TR_8, is supplied by a negative feedback network from the output pin 9. The output from the long-tail pair is amplified by TR_6 and TR_7 and passed to the phase splitting stage TR_{17} and TR_{18}. The emitter of TR_{17} and TR_{18} are biased by diode and further transistor networks.

The push-pull output stage contains four transistors, TR_{22} to TR_{25}. Transistors TR_{22} and TR_{24} amplify the positive-going part of the signal, and transistors TR_{23} and TR_{25} amplify the negative-going part.

The a.f. output is taken from pin 9 of the IC to the 4Ω speaker.

Performance

The signal-to-noise characteristic for a.m. reception is shown in Fig. 55. The a.f. voltage across the volume control is measured with the slider at the lower end, and the r.f. voltage is measured at the base of the mixer/oscillator transistor TR_{26}. The signal-to-noise characteristic for f.m. operation is shown in Fig. 56. The a.f. voltage is again measured across the volume control with the slider at the lower end, but the r.f. voltage is measured at the input to the dummy aerial shown in the circuit of Fig. 57.

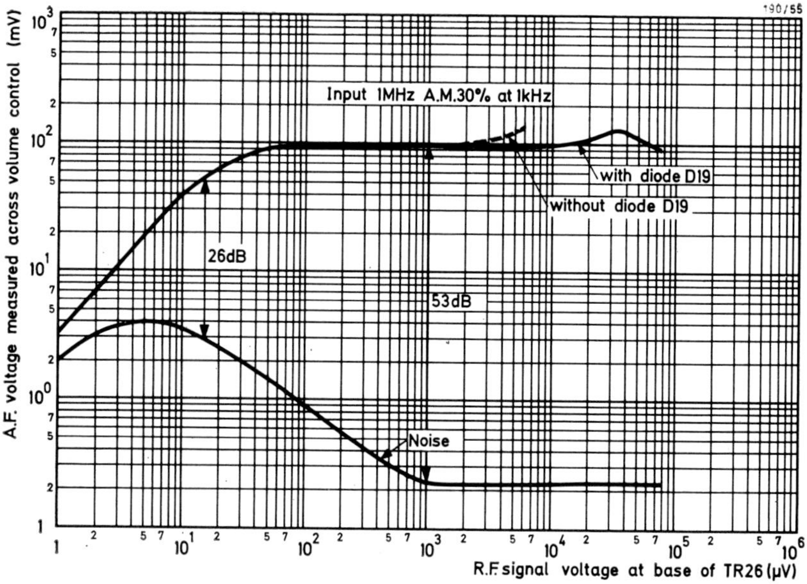

Fig. 55—Signal-to-noise characteristic for a.m. reception

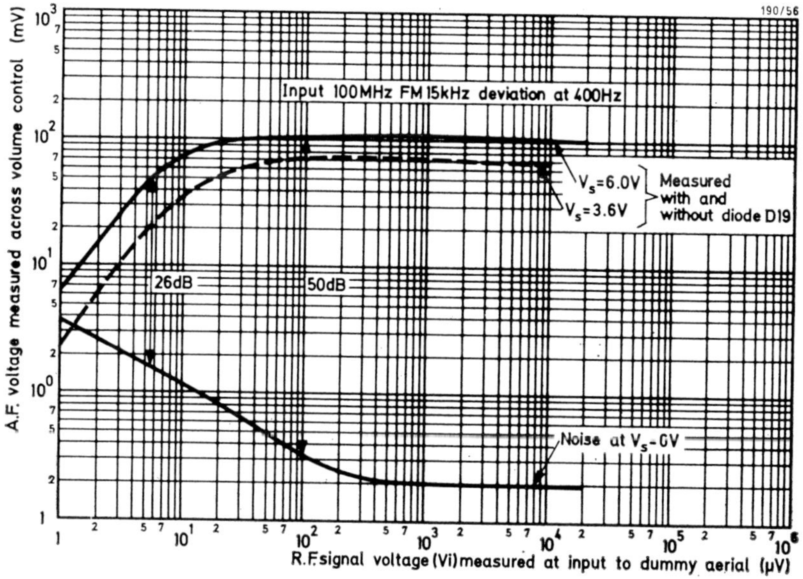

Fig. 56—Signal-to-noise characteristic for f.m. reception

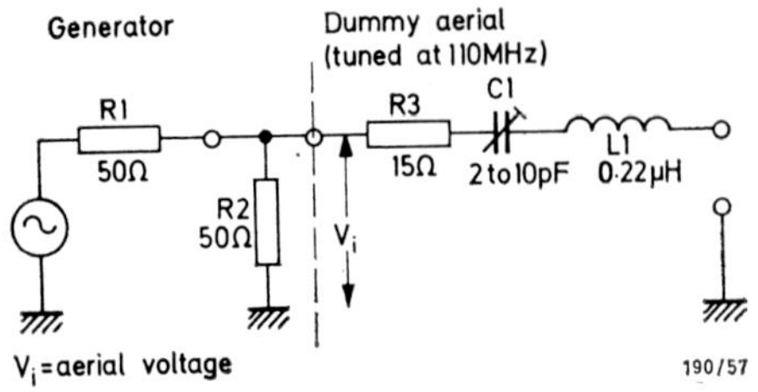

Fig. 57—Dummy aerial for f.m. measurements

COMPONENT LISTS FOR CIRCUITS IN CHAPTER 4

1W Audio Amplifier (page 35)

Transistors

Circuit reference	Mullard type
TR_1	AC127
TR_2	AC128
TR_3	AC128
TR_4	AC127

Resistors

Mullard CR25 series (0·3W) ±5% except where stated

Circuit reference	Value
R_1	2·2kΩ
R_2	15kΩ
R_3	15kΩ
R_4	2·7Ω
R_5	2·2kΩ
R_6	1·5kΩ
R_7	510Ω
R_8 thermistor VA1040	130Ω
R_9 preset potentiometer	200Ω
R_{10}	39Ω

Capacitors

Circuit reference	Value	Description	Mullard type
C_1	10μF	electrolytic 16V	C426AR/E10
C_2	125μF	electrolytic 10V	C426AR/D125
C_3	400μF	electrolytic 6·4V	C437AR/D400
C_4	4·7nF	polyester	C296AC/A4K7
C_5	320μF	electrolytic 2·5V	C426AR/A320
C_6	320μF	electrolytic 6·4V	C426AR/C320

Heatsinks

For output transistor TR_3 (AC128) a cooling clip is required. For output transistor TR_4 (AC127) a cooling clip mounted on a 1·5mm (16 gauge) bright aluminium heatsink with an area of 12·5cm^2 is required.

2W Audio Amplifier (page 37)

Integrated circuit

Mullard TCA160

Resistors

Mullard CR25 series (0·3W) ±10%

Circuit reference	Value
R_1	47Ω
R_2	5·1kΩ

Capacitors

Circuit reference	Value	Description	Mullard type
C_1	3·3nF (typical)	polyester film/foil	347 61332
C_2	1·6μF	electrolytic 25V	C426AS/F1·6
C_3	32μF	electrolytic 10V	C426AR/D32
C_4	125μF	electrolytic 16V	C426AR/E125
C_5	640μF	electrolytic 10V	C437AR/D640
C_6	150nF	metallised polyester	C280AE/P150K
C_7	250μF	electrolytic 16V	C437AR/E250

3W Audio Amplifier (page 39)

Transistors and Rectifier

Circuit reference	Mullard type
TR_1	BC148 or BC108
TR_2	AC128
TR_3	AC176
TR_4	AC128
Full-wave bridge rectifier	BY164

Resistors

Mullard CR25 series (0·3W) ±5% except R_8, R_9, R_{10} which are CR37 (0·5W) ±5%

Circuit reference	Value	Circuit reference	Value
R_1	39kΩ	R_6	2·2kΩ
R_2	120kΩ	R_7	15Ω
R_3	47kΩ	R_8	560Ω
R_4	390Ω	R_9	2·2Ω
R_5	22Ω	R_{10}	2·2Ω

Capacitors

Circuit reference	Value	Description	Mullard type
C_1	220nF	polyester	C296AA/A220K
C_2	25μF	electrolytic 25V	C426AR/F25
C_3	250μF	electrolytic 16V	C437AR/E250
C_4	3·3nF	polyester	C296AC/A3K3
C_5	400μF	electrolytic 16V	C437AR/E400
C_6 (mono)	1250μF	electrolytic 25V	C431BR/F1250
C_6 (stereo)	2000μF	electrolytic 25V	C431BR/F2000

Mains Transformer

The transformer used in this circuit may be obtained from the following manufacturers under the type numbers given:

Colne Electric Limited	20040
Drake Transformers Limited	1471-136
Gardners Transformers Limited	GR 97180
Parmeko Limited	P 3201

Heatsinks

The driver and output transistors should be mounted in clips having good thermal contact and fixed to heatsinks of 1·5mm (16 gauge) bright aluminium. The area required for the driver is $12cm^2$ and for each output transistor $35cm^2$.

10W Audio Amplifier (page 41)

Transistors and Rectifier

Circuit reference	Mullard type
TR_1	BC148 or BC108
TR_2	AC128
TR_3	AD161
TR_4	AD162
Rectifier bridge	4×BY126

Resistors

Mullard CR25 series (0·3W) ±5% except R_{11}, R_{12} which are CR37 (0·5W) ±5%, and R_{10} which is 1W

Circuit reference	Value	Circuit reference		Value
R_1	15kΩ	R_7		12Ω
R_2	68kΩ	R_8	thermistor VA1100	15Ω
R_3	100kΩ	R_9	preset potentiometer	22Ω
R_4	10Ω	R_{10}		220Ω
R_5	470Ω	R_{11}		0·5Ω
R_6	680Ω	R_{12}		0·5Ω

Capacitors

Circuit reference	Value	Description	Mullard type
C_1	50μF	electrolytic 40V	C426AR/G50
C_2	25μF	electrolytic 25V	C426AR/F25
C_3	640μF	electrolytic 16V	C437AR/E640
C_4	500μF	electrolytic 2·5V	C426AR/A500
C_5	1000μF	electrolytic 16V	C431BR/E1000
C_6	15nF	polyester film foil	C296AA/A15K
C_7	4000μF	electrolytic 25V	C431BR/F4000

Mains Transformer

The transformer used in this circuit may be obtained from the following manufacturers under the type numbers given:

Colne Electric Limited	20041
Drake Transformers Limited	294-139
Gardners Transformers Limited	GR 97181
Parmeko Limited	P 3202

Heatsinks

Transistors TR_2, TR_3 and TR_4 should be mounted on a common 1·5mm (16 gauge) bright aluminium heatsink with a minimum area of 100cm^2 or blackened aluminium with an area of 60cm^2. Transistor TR_2 has an isolated can, but TR_3 and TR_4 should both be mounted on mica washers.

Pre-amplifier (page 44)

Transistors

Circuit reference	Mullard type
TR_1	BC149 or BC109
TR_2	BC148 or BC108

Resistors

Mullard CR25 series (0·3W) ±5% except where stated

Circuit reference		Value
R_1		see text
R_2		1·8MΩ
R_3		270kΩ
R_4		2MΩ
R_5		47kΩ
R_6		47kΩ
R_7	linear potentiometer	470kΩ
R_8	linear potentiometer	220kΩ
R_9		47kΩ
R_{10}		150kΩ
R_{11}		150kΩ
R_{12}		47kΩ
R_{13}		22kΩ
R_{14}		2·2kΩ
R_{15}		15kΩ
R_{16}	linear potentiometer	47kΩ
R_{17}	logarithmic potentiometer	22kΩ

Capacitors

Circuit reference	Value	Description	Mullard type
C_1	10nF	polyester	C296AA/A10K
C_2	10μF	electrolytic 16V	C426AR/E10
C_3	40μF	electrolytic 16V	C426AR/E40
C_4	1·5nF	polyester	C296AC/A1K5
C_5	6·8nF	polyester	C296AC/A6K8
C_6	6·8nF	polyester	C296AC/A6K8
C_7	1·5nF	polyester	C296AC/A1K5
C_8	10μF	electrolytic 16V	C426AR/E10
C_9	10μF	electrolytic 16V	C426AR/E10
C_{10}	10μF	electrolytic 16V	C426AR/E10
C_{11}	10pF	polystyrene	
C_{12}	see text		

3W Record-player Amplifier

Components for Basic Version (page 46)

Transistors and Rectifier

Circuit reference	Mullard type
TR_1	BC108
TR_2	BC107
TR_3	AC128
TR_4	AC176
Full-wave bridge rectifier	BY164

Resistors

Mullard CR25 series (0·3W) ±5% except R_6 and R_7 which are CR37 (0·5W) ±5%

Circuit reference	Value	Circuit reference	Value
R_1 logarithmic potentiometer	2MΩ	R_7	150Ω
		R_8	2·2Ω
R_2	120kΩ	R_9	2·2Ω
R_3	470kΩ	R_{10} logarithmic potentiometer	250kΩ
R_4	390kΩ		
R_5	15Ω	R_{11}	1·2kΩ
R_6	470Ω	R_{12}	15kΩ

Capacitors

Circuit reference	Value	Description	Mullard type
C_1	100nF	polyester	C296AA/A100K
C_2	100nF	polyester	C296AA/A100K
C_3	40μF	electrolytic 16V	C426AR/E40
C_4	10nF	polyester	C296AA/A10K
C_5	680pF	polystyrene	C295AC/680E
C_6	250μF	electrolytic 16V	C437AR/E250
C_7	640μF	electrolytic 25V	C437AR/F640

Mains Transformer

The transformer used in this circuit may be obtained from the following manufacturers under the type numbers given:

Colne Electric Limited	20042
Drake Transformers Limited	351-245
Gardners Transformers Limited	GR 97182
Parmeko Limited	P 3203

Components for Version including Comprehensive Tone Control (page 49)

Transistors and Rectifier

Circuit reference	Mullard type
TR_1	BC109
TR_2	BC107
TR_3	AC128
TR_4	AC176
Full-wave bridge rectifier	BY164

Resistors

Mullard CR25 series (0·3W) ±5% except R_6 and R_7 which are CR37 (0·5W) ±5%

Circuit reference	Value	Circuit reference	Valve
R_1 logarithmic potentiometer	2MΩ	R_8	2·2Ω
		R_9	2·2Ω
R_2	220kΩ	R_{10} reverse logarithmic potentiometer	100kΩ
R_3	1MΩ		
R_4	820kΩ	R_{11}	1·2kΩ
R_5	15Ω	R_{12}	15kΩ
R_6	470Ω	R_{13} reverse logarithmic potentiometer	100kΩ
R_7	150Ω		

Capacitors

Circuit reference	Value	Description	Mullard type
C_1	2·5μF	electrolytic 1·5V	
C_2	100nF	polyester	C296AA/A100K
C_3	40μF	electrolytic 16V	C426AR/E40
C_4	4·7nF	polyester	C296AC/A4K7
C_5	680pF	polystyrene	C295AC/680E
C_6	250μF	electrolytic 16V	C437AR/E250
C_7	640μF	electrolytic 25V	C437AR/F640
C_8	47nF	polyester	C296AA/A47K
C_9	6·4μF	electrolytic 25V	C426AR/F6·4
C_{10}	22nF	polyester	C296AA/A22K
C_{11}	220nF	polyester	C296AA/A220K

Mains Transformer

For suppliers of the transformer, see page 74.

Heatsinks

Each output transistor should be mounted on a 1·5mm (16 gauge) bright aluminium heatsink with an area of 30cm^2. In the circuit shown the driver transistor does not require a heatsink. However, an increase in output power (3W at clipping) may be obtained, by fitting a heat clip on the driver transistor and reducing the value of R_6 to 330Ω to give a higher drive current, with a 12Ω load. (Alternatively, a larger driver transistor, the BFY51, may be used with the same reduction in the value of R_6 and a 12Ω load.)

A.M. Receiver (page 52)

Transistors and Diode

See note on page 51

Circuit reference	Mullard type
TR_1	BF194
TR_2	BF195
TR_3	BF195
D_1	OA90

Resistors

Mullard CR16 series (0·2W) ±5% except where stated

Circuit reference	Value	Circuit reference	Value
R_1	18kΩ	R_5	330kΩ
R_2	15kΩ	R_6	120kΩ
R_3	2·7kΩ	R_7	27kΩ
R_4	47kΩ	R_8 logarithmic carbon potentiometer	10kΩ

Capacitors

Circuit reference	Value	Description	Mullard type
C_1 } C_5 }		foil dielectric variable capacitor	AC0039
C_2	10nF	polyester	C296AA/A10K
C_3	220pF	ceramic plate	C333CH/C220E
C_4	22nF	polyester	C296AA/A22K
C_6	1·6μF	electrolytic 25V	
C_7	220pF	ceramic plate	C333CH/C220E
C_8	100nF	polyester	C296AA/A100K
C_9	220pF	ceramic plate	C333CH/C220E
C_{10}	10nF	polyester	C296AA/A10K

Transformers

Aerial transformer—T_1

L_1	80 turns of 12×46 s.w.g. stranded wire; close-wound to fit Mullard 9·5mm diameter ferrite rod, type FX1247
L_2	3½ turns of 12×46 s.w.g. stranded wire; interwound with earth end of L_1

Oscillator transformer—T_2

L_3	10 turns of 3×46 s.w.g. stranded wire
L_4	2 turns of 3×46 s.w.g. stranded wire
L_5	100 turns of 3×46 s.w.g. stranded wire; wave-wound on 4·8mm former

First i.f. transformer—T_3

L_6+L_7	42+108 turns of 3×46 s.w.g. stranded wire
L_8	2 turns of 3×46 s.w.g. stranded wire; wave-wound on 4·8mm former; $Q_O \simeq Q_L = 80$ when L_6+L_7 is shunted with 330kΩ resistor

Second i.f. transformer—T_4

L_9+L_{10}	80+70 turns of 3×46 s.w.g. stranded wire
L_{11}	9 turns of 3×46 s.w.g. stranded wire; wound on 4·8mm former; $Q_O = 120$, $Q_L = 95$

Third i.f. transformer—T_5

$L_{12}+L_{13}$	51+99 turns of 3×46 s.w.g. stranded wire
L_{14}	32 turns of 3×46 s.w.g. stranded wire; wound on 4·8mm former; $Q_O = 120$, $Q_L = 40$

A.M./F.M. Receiver (pages 56 and 57)

Transformers and Coils

Except where otherwise stated, all transformers and coils are wound on 4mm formers with ferrite frames and slugs.

F.M. aerial transformer—T_1

$L_1 + L_2$ — $2+2$ turns of 24 s.w.g. enamelled-copper wire; wound on ferrite twin bead

First f.m. i.f. transformer—T_2

$L_6 + L_7$ — $14+14$ turns of 36 s.w.g. enamelled-copper wire; wound on a 4·4mm diameter former
L_7 is critically coupled to L_6 when L_7 is loaded with 600Ω at the capacitive tap

Medium-wave aerial transformer—T_3

L_8 — 80 turns of 12×46 s.w.g. silk-covered enamelled-copper wire; wound on Mullard ferrite rod, type FX1247 ($203 \times 9{\cdot}5$mm); $Q_O = 200$

L_9 — $3\frac{1}{2}$ turns of 12×46 s.w.g. silk-covered enamelled-copper wire; interwound with earth end of L_8

A.M. oscillator transformer—T_4

L_{10} (primary) — $119\frac{7}{8}$ turns of 6×48 s.w.g. enamelled-copper wire; $Q_O = 145$

L_{11} (collector) — $9\frac{1}{2}$ turns of 6×48 s.w.g. enamelled-copper wire

L_{12} (emitter) — $1\frac{7}{8}$ turns of 6×48 s.w.g. enamelled-copper wire;
wound on 'baby' coil former;

First a.m. i.f. transformer T_5

L_{13} (primary) — 130 turns of 3×46 s.w.g. enamelled-copper wire; tapped at 41 turns from the earth end; tap impedance$=6{\cdot}5$kΩ; $Q_O = 75$, $Q_L = 70$

L_{14} (secondary) — $5\frac{1}{4}$ turns of 3×46 s.w.g. enamelled-copper wire

Turns ratio, primary tap: secondary$=7{\cdot}8:1$

Second f.m. i.f. transformer—T_6

L_{15} (primary) — 8 turns of 36 s.w.g. enamelled-copper wire; $Q_O = Q_L = 53$

L_{16} (secondary) — 6 turns of 36 s.w.g. enamelled-copper wire tapped at $1\frac{3}{4}$ turns from earth end; unloaded tap impedance$=118$Ω; $Q_O = 53$, $Q_L = 50$
L_{16} is critically coupled when secondary tap is loaded with $2{\cdot}2$kΩ

Turns ratio, primary:secondary tap$=4{\cdot}5:1$

Second a.m. i.f. transformer—T_7

L_{17} (primary) — 110 turns of 3×46 s.w.g. enamelled-copper wire tapped at 73 turns from earth end; tap impedance$=35{\cdot}6$kΩ; $Q_O = 98$, $Q_L = 70$

L_{18} (secondary) — 10 turns of 3×46 s.w.g. enamelled-copper wire

Turns ratio, primary tap:secondary$=6{\cdot}9:1$

Third f.m. i.f. transformer—T_8

L_{19} (primary) — As L_{15} (T_6)

L_{20} (secondary) — As L_{16} (T_6)

Third a.m. i.f. transformer—T_9

L_{21} (primary)	110 turns of 3×46 s.w.g. enamelled-copper wire tapped at 48 turns from earth end; tap impedance=21·6kΩ; Q_O=140, Q_L=50
L_{22} (secondary)	31 turns of 3×46 s.w.g. enamelled-copper wire

Turns ratio, primary tap:secondary=1·53:1

F.M. ratio detector transformer—T_{10}

L_{23} (primary)	8 turns of 34 s.w.g. enamelled-copper wire; Q_O=90, Q_L=58
L_{24} (secondary)	2×7 turns of 36 s.w.g. enamelled-copper wire; Q_O=75
L_{25} (tertiary)	3½ turns of 36 s.w.g. enamelled-copper wire; overwound on primary at earth end. Primary to secondary coupling is 0·7 with secondary loaded with 10kΩ and tertiary open-circuited.
Coil L_3	3½ turns of 20 s.w.g. tinned-copper wire; spaced one diameter and wound on a 6mm diameter former
Coil L_4	12 turns of 28 s.w.g. enamelled-copper wire; wound on a 4·4mm former with ferrite core
Coil L_5	3½ turns of 20 s.w.g. tinned-copper wire; spaced two diameters and wound on a 6mm diameter former; tapped two turns from the earth end

Transistors and Diodes

Circuit reference	Mullard type	Circuit reference	Mullard type
TR_1	BF195	D_1	OA90
TR_2	BF195	D_2	AA119
TR_3	BF195	D_3	AA119
TR_4	BF194		
TR_5	BF194		

Resistors

Mullard CR16 series (0·2W) ±5% except where stated

Circuit reference	Value	Circuit reference	Value
R_1	1kΩ	R_{14}	39kΩ
R_2	5·6kΩ	R_{15}	15kΩ
R_3	18kΩ	R_{16}	4·7kΩ
R_4	680Ω	R_{17}	100Ω
R_5	5·6kΩ	R_{18}	1·2kΩ
R_6	18kΩ	R_{19}	10Ω
R_7	27kΩ	R_{20}	1kΩ
R_8	39kΩ	R_{21}	470Ω
R_9	100Ω	R_{22}	20kΩ
R_{10}	3·9kΩ	R_{23} CR16 ±10%	1·5MΩ
R_{11}	120kΩ	R_{24} logarithmic potentiometer	20kΩ
R_{12}	3·9kΩ		
R_{13}	100Ω	R_{25}	330Ω

Capacitors

Circuit reference	Value	Description	Mullard type
C_1	1nF	polyester	C296AC/A1K
C_2	12pF	ceramic plate	C333CB/L12E
C_3	1nF	polyester	C296AC/A1K
C_4	2 to 12pF	trimmer	
C_5	2 to 13·6pF	ganged tuning capacitor	
C_{13}	2 to 13·6pF	ganged tuning capacitor	
C_6	33pF	ceramic plate	C333CC/C33E
C_7	4·7pF	ceramic plate	C333CB/L4E7
C_8	270pF	polystyrene	
C_9	39pF	ceramic plate	C333CC/C39E
C_{10}	1nF	polyester	C296AC/A1K
C_{11}	4·7pF	ceramic plate	C333CB/L4E7
C_{12}	2 to 12pF	trimmer	
C_{14}	33pF	ceramic plate	C333CC/C33E
C_{15}	68pF	ceramic plate	C333CH/C68E
C_{16}	1nF	polyester	C296AC/A1K
C_{17}	68pF	ceramic plate	C333CH/C68E
C_{18}	1nF	polyester	C296AC/A1K
C_{19}	47nF	metallised foil	C280AE/P47K
C_{20}, C_{21}	ganged tuning capacitor, Plessey 'W' type		
C_{22}	22nF	metallised foil	C280AE/P22K
C_{23}	390pF	polystyrene	
C_{24}	180pF	ceramic plate	C333CH/C180E
C_{25}	2·5μF	electrolytic 16V	C426AR/H2·5
C_{26}	470pF	polystyrene	
C_{27}	10nF	metallised foil	C280AE/P10K
C_{28}	390pF	polystyrene	
C_{29}	180pF	ceramic plate	C333CH/C180E
C_{30}	470pF	polystyrene	
C_{31}	47nF	metallised foil	C280AE/P47K
C_{32}	390pF	polystyrene	
C_{33}	180pF	ceramic plate	C333CH/C180E
C_{34}	47nF	metallised foil	C280AE/P47K
C_{35}	56pF	ceramic plate	C333CH/C56E
C_{36}	10nF	metallised foil	C280AE/P10K
C_{37}	330pF	polystyrene	
C_{38}	330pF	polystyrene	
C_{39}	4·7nF	polyester	C296AC/A4K7
C_{40}	47nF	metallised foil	C280AE/P47K
C_{41}	10μF	electrolytic 16V	C426AR/E10
C_{42}	2·5μF	electrolytic 16V	C426AR/H2·5
C_{43}	10μF	electrolytic 16V	C426AR/E10

F.M. I.F. and Detector Stages (page 61)

Semiconductor Devices

Circuit reference	Mullard type
Integrated circuit	TBA750
TR_1	BF194
TR_2	AC127
D_1	AA119
D_2	AA119

Resistors

Mullard CR16 series (0·2W) ±5%

Circuit reference	Value
R_1	680Ω
R_2	5·6kΩ
R_3	220Ω
R_4	1kΩ
R_5	3kΩ
R_6	3kΩ
R_7	10kΩ
R_8	12kΩ
R_9	820Ω
R_{10}	1kΩ

Capacitors

Circuit reference	Value	Description	Mullard type
C_1	10nF	metallised foil	C280AE/P10K
C_2	390pF	polystyrene	
C_3	10nF	metallised foil	C280AE/P10K
C_4	100nF	metallised foil	C280AE/P100K
C_5	390pF	polystyrene	
C_6	22nF	metallised foil	C280AE/P22K
C_7	22nF	metallised foil	C280AE/P22K
C_8	10nF	metallised foil	C280AE/P10K
C_9	100pF	ceramic plate	C333CC/C100E
C_{10}	330pF	ceramic plate	C333CH/C330E
C_{11}	22pF	ceramic plate	C333CC/C22E
C_{12}	1nF	polystyrene	C295AC/D1K
C_{13}	10nF	metallised foil	C280AE/P10K
C_{14}	22pF	ceramic plate	C333CC/C22E
C_{15}	22nF	metallised foil	C280AE/P22K
C_{16}	10μF	electrolytic 16V	C426AR/E10

A.M./F.M. Receiver (pages 64 and 65)

Semiconductor Devices

Circuit reference	Mullard type	Circuit reference	Mullard type
Integrated circuit	TBA 700		
TR_{26}	BF195	D_{21}	AA119
D_{19}	AA119	D_{22}	AA119
D_{20}	AA119		

(Transistors TR_1 to TR_{25}, and diodes D_1 to D_{18}, are inside the integrated circuit)

Capacitors

Circuit reference	Value	Description	Mullard Type
C_1	1nF	ceramic plate	629 02102
C_2	100nF	metallised foil	C280AE/P100K
C_3	82pF	ceramic plate	C333CC/C82E
C_4	1nF	ceramic plate	629 02102
C_5	2 to 10pF	trimmer	
C_6	10pF	ceramic plate	C333CB/C10E
C_7	2 to 22pF	trimmer	
C_8	120pF	ceramic plate	C333CC/C120E
C_9, C_{21}	5 to 385pF	tuning capacitor	
C_{10}	150pF	ceramic plate	C333CC/C150E
C_{11}	150pF	ceramic plate	C333CC/C150E
C_{12}	1nF	ceramic plate	629 02102
C_{13}	1nF	ceramic plate	629 02102
C_{14}	180pF	ceramic plate	C333CH/C180E
C_{15}	22nF	metallised foil	C280AE/P22K
C_{16}	2 to 10pF	trimmer	
C_{17}	10pF	ceramic plate	C333CC/C10E
C_{18}	5·6nF	polystyrene	C295AH/D5K6
C_{19}	2·2nF	ceramic plate	629 02222
C_{20}	470pF	polystyrene	
C_{22}	410pF	polystyrene	
C_{23}	22pF	ceramic plate	C333CC/C22E
C_{24}	2 to 22pF	trimmer	
C_{25}	22nF	metallised foil	C280AE/P22K
C_{26}	150pF	ceramic plate	C333CC/C150E
C_{27}	1μF	electrolytic 40V	C426AS/G1
C_{28}	22pF	ceramic plate	C333CC/C22E
C_{29}	3·3nF	polystyrene	C295AA/D3K3
C_{30}	1nF	polystyrene	C295AC/D1K

Circuit reference	Value	Description	Mullard Type
C_{31}	100nF	metallised foil	C280AE/P100K
C_{32}	150pF	ceramic plate	C333CC/C150E
C_{33}	1nF	polystyrene	C295AC/D1K
C_{34}	10μF	electrolytic 16V	C426AR/E10
C_{35}	2·2nF	ceramic plate	629 02222
C_{36}	150pF	ceramic plate	C333CC/C150E
C_{37}	150pF	ceramic plate	C333CC/C150E
C_{38}	22nF	metallised foil	C280AE/P22K
C_{39}	22nF	metallised foil	C280AE/P22K
C_{40}	39pF	ceramic plate	C333CC/C39E
C_{41}	64μF	electrolytic 10V	C426AR/D64
C_{42}	22nF	metallised foil	C280AE/P22K
C_{43}	4·7nF	ceramic plate	629 02472
C_{44}	4·7nF	ceramic plate	629 02472
C_{45}	10μF	electrolytic 16V	C426AR/E10
C_{46}	82pF	ceramic plate	C333CC/C82E
C_{47}	2·2nF	ceramic plate	629 02222
C_{48}	1μF	electrolytic 40V	C426AS/G1
C_{49}	220pF	ceramic plate	C333CH/C220E
C_{50}	220pF	ceramic plate	C333CH/C220E
C_{51}	680pF	ceramic plate	630 02681
C_{52}	10μF	electrolytic 16V	C426AR/E10
C_{53}	10μF	electrolytic 16V	C426AR/E10
C_{54}	10μF	electrolytic 16V	C426AR/E10
C_{55}	1·8nF	ceramic plate	630 02182
C_{56}	470pF	ceramic plate	630 02471
C_{57}	2·2nF	ceramic plate	629 02222
C_{58}	1μF	electrolytic 40V	C426AS/G1
C_{59}	330μF	electrolytic 10V	016 14331
C_{60}	100pF	metallised foil	C280AE/P100K
C_{61}	200μF	electrolytic 10V	C426AR/D200
C_{62}	330μF	electrolytic 10V	016 14331

Resistors

Mullard CR16 series (0·2W) ±5% except where stated

Circuit reference	Value	
R_{12}	100Ω	
R_{13}	220kΩ	
R_{14}	150kΩ	
R_{15}	3·3kΩ	
R_{16}	680Ω	
R_{17}	18Ω	
R_{18}	220Ω	
R_{19}	200Ω	
R_{20}	220Ω	
R_{21}	820Ω	
R_{22}	220Ω	
R_{23}	100Ω	
R_{24}	100kΩ	
R_{25}	1kΩ	
R_{26}	22kΩ	
R_{27}	2·2kΩ	
R_{28}	10kΩ	logarithmic potentiometer
R_{29}	390Ω	
R_{30}	470Ω	
R_{31}	100Ω	
R_{32}	10kΩ	
R_{33}	10kΩ	
R_{34}	39kΩ	
R_{35}	33kΩ	
R_{36}	150Ω	for 4Ω speaker
	68Ω	for 8Ω speaker
R_{37}	150Ω	
R_{38}	15kΩ	
R_{39}	3·9kΩ	
R_{40}	3·3kΩ	
R_{41}	5·6Ω	
R_{42}	68Ω	

(Resistors R_1 to R_{11} are inside the integrated circuit)

Transformers and Coils

A.M. S.W. aerial coil L_1

N_1+N_2	$13+0{\cdot}5$ turns of $16\times0{\cdot}03$mm silk-covered enamelled-copper wire; $Q_0=100$.

A.M. M.W. aerial coil L_2

	50 turns of $36\times0{\cdot}03$mm silk-covered enamelled-copper wire; wound on ferrite rod aerial 200×10mm; $Q_0=100$.

A.M. L.W. aerial coil L_3

N_1+N_2	$150+12{\cdot}5$ turns of $24\times0{\cdot}04$mm silk-covered enamelled-copper wire; wound on ferrite rod aerial, direction of winding of L_3 N_2 the same as L_2; $Q_0=22$.

First f.m. i.f. bandpass filter

Primary L_4

N_1+N_2	$16+1{\cdot}5$ turns of $6\times0{\cdot}03$mm silk-covered enamelled-copper wire; $Q_0=90$.

Secondary L_5

N_1+N_2	$17+3$ turns of $3\times0{\cdot}03$mm silk-covered enamelled-copper wire; $Q_0=55$.

A.M. S.W. oscillator coil L_6

$N_1+N_2+N_3$	$13+1+4$ turns of $16\times0{\cdot}03$mm silk-covered enamelled-copper wire; $Q_0=80$.

A.M. L.W./M.W. oscillator coil L_7

$N_1+N_2+N_3$	$90+3+10$ turns of $16\times0{\cdot}03$mm silk-covered enamelled-copper wire; $Q_0=200$.

First f.m. single-tuned filter L_8

N_1+N_2	$17+3$ turns of $3\times0{\cdot}03$mm silk-covered enamelled-copper wire; tuning capacitor inside sreening can; $Q_0=55$.

First a.m. bandpass filter

Primary L_9

N_1	50 turns of $16\times0{\cdot}03$mm silk-covered enamelled-copper wire; $Q_0=90$.

Secondary L_{10}

N_1+N_2	$90+5$ turns of $6\times0{\cdot}03$mm silk-covered enamelled-copper wire; $Q_0=80$.

Second f.m. single-tuned filter L_{11}

N_1+N_2	$17+3$ turns of $3\times0{\cdot}03$mm silk-covered enamelled-copper wire; tuning capacitor inside screening can; $Q_0=55$.

First a.m. single-tuned filter L_{12}

N_1+N_2	$90+3$ turns of $6\times0{\cdot}03$mm silk-covered enamelled-copper wire; $Q_0=80$.

Ratio detector transformer, primary L_{13}

N_1+N_2	$12+6$ turns of $12\times0{\cdot}03$mm silk-covered enamelled-copper wire; tuning capacitor inside screening can; $Q_0=95$.

A.M. detector coil L_{14}

$N_1+N_2+N_3$	185+60+60 turns of 6×0·03mm silk-covered enamelled copper wire; tuning capacitor inside screening can; $Q_0=130$.

Ratio detector secondary L_{15}

$N_1+N_2+N_3$	8+8+2 turns of 12×0·03mm silk-covered enamelled-copper wire; bifilar wound with tuning capacitor inside screening can; $Q_0=110$.

CHAPTER 5

TAPE RECORDERS

In this chapter, a 4W tape recorder amplifier and a tape pre-amplifier are described, and the modifications necessary for stereo operation are considered. An automatic gain control circuit for use with tape recorders is also included.

The amplifier and pre-amplifier circuits are designed around the Marriott type X/RPS/36 recording head and X/ES/11 erase head; they will not operate with heads of different inductance and bias requirements.

Component lists for the circuits are given at the end of the chapter.

4W TAPE RECORDER

The 4W tape recorder shown in Fig. 59 uses silicon transistors in the first three stages to obtain high sensitivity. The output stage uses a complementary matched pair, transistors AD161 (n-p-n) and AD162 (p-n-p), which are also used for the oscillator during recording. Equalisation is provided for tape speeds of $7\frac{1}{2}$in/s (19·1cm/s), $3\frac{3}{4}$in/s (9·5cm/s), $1\frac{7}{8}$in/s (4·8cm/s) and $\frac{15}{16}$in/s (2·4cm/s), which conforms to C.C.I.R. standards where appropriate.

Circuit Description

Playback

The BC149 or BC109 has been chosen for the input stage because it is particularly suitable for use in audio input stages where low noise is essential. Its collector is directly connected to the base of TR_2; its emitter is connected via an RC network to the collector of TR_2. This RC network provides a path for negative feedback by means of which equalisation is effected. The frequency response curves on playback are shown in Fig. 60.

A d.c. feedback path is also provided from the emitter of TR_2 to the base of TR_1. The d.c. feedback along this path stabilises the working points of TR_1 and TR_2.

The other transistors in the tape recorder form a conventional amplifier. The complementary pair, TR_5 and TR_6, is driven by TR_4, which is itself driven by the high-gain BC148 or BC108 transistor. Potentiometer R_{23} is adjusted to give a quiescent output-stage current of 5mA.

Record

Different input circuits are needed to match the different signal sources. For example, the input circuit for a crystal microphone contains a shunt capacitor, C_1, and that for a radio receiver is a potential divider. A moving coil microphone, however, can be connected directly because the amplifier has an input resistance of 200kΩ. During recording, TR_5 and TR_6 operate as an oscillator at 50kHz, the erase head being used as an oscillator coil. The capacitive taps, C_{20}, C_{21} and C_{22}, also provide the necessary conditions for sustained oscillation.

The recording level indicator gives a reading proportional to the peak-to-peak voltage at the collector of TR_4. The full recording level with a sinewave is 4V r.m.s., which produces a current of 95μA through the meter and 110μA r.m.s. through the recording head.

During recording, the equalisation circuits connected to switch S_C are not used; feedback is via the path C_3, R_6 and R_7, which has been designed to give treble boost. The frequency response curves on record are shown in Fig. 61.

The recording current is taken from the collector of TR_4, and the recording head is a Marriott type X/RPS/36* with the following characteristics:

Quarter-track head:	
Gap	0·0001in
Width	0·043in
Spacing	0·136in
Inductance at 1kHz	70mH
Bias at 50kHz	>11V
Recording current for 2dB below 2% distortion	110μA
Output at 1kHz using E.M.I. test tape T.B.T.I.	0·44mV
Output at 2kHz for 110μA recording current	1·07mV

Power Supply

The circuit of a simple power supply suitable for use with the tape recorder is shown in Fig. 58. It contains a mains transformer with a secondary output of 17V. This is rectified by means of a bridge rectifier, type BY164, and smoothed by means of a 640μF capacitor.

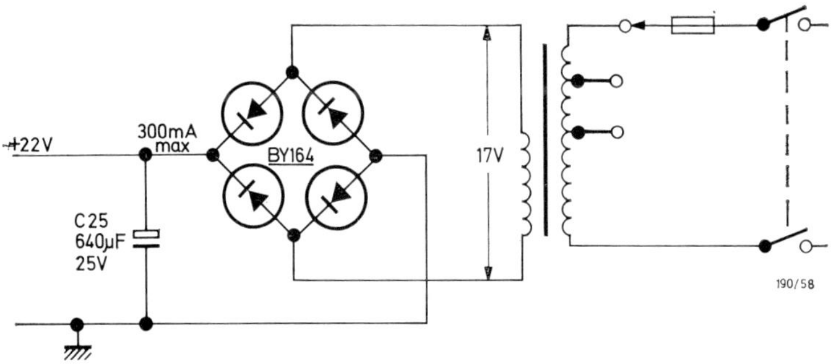

Fig. 58—Power supply circuit for a 4W tape recorder

*Record/playback heads with a larger inductance are not suitable because they would require a larger bias voltage.

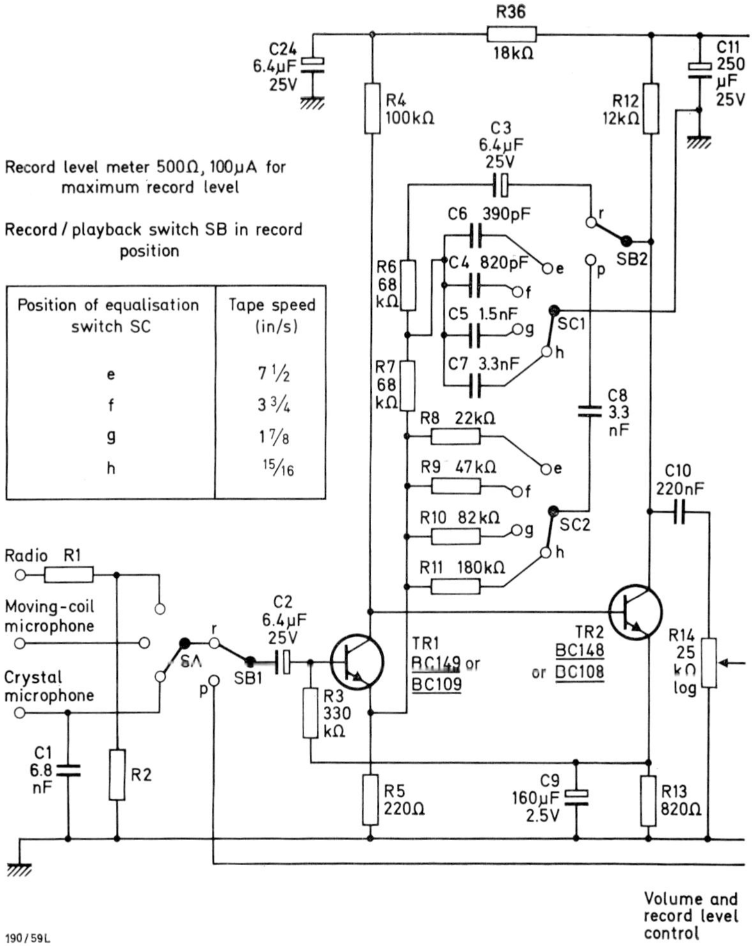

Position of equalisation switch SC	Tape speed (in/s)
e	7 1/2
f	3 3/4
g	1 7/8
h	15/16

Fig. 59—4W tape recorder circuit

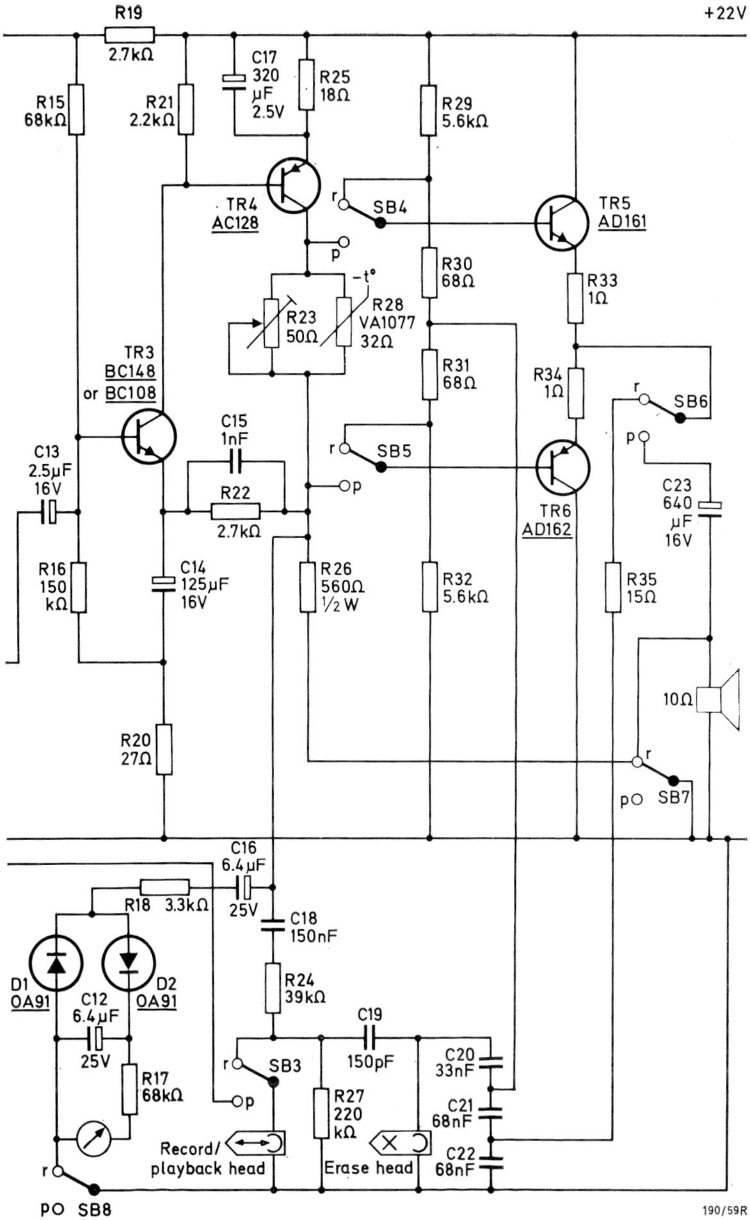
R19
2.7kΩ
+22V
R15
68kΩ
R21
2.2kΩ
C17
320
μF
2.5V
R25
18Ω
R29
5.6kΩ
TR4
AC128
SB4
TR5
AD161
R30
68Ω
R33
1Ω
R23
50Ω
R28
VA1077
32Ω
TR3
BC148
or BC108
R31
68Ω
R34
1Ω
SB6
C15
1nF
SB5
C13
2.5μF
16V
R22
2.7kΩ
TR6
AD162
C23
640
μF
16V
C14
125μF
16V
R16
150
kΩ
R26
560Ω
½ W
R32
5.6kΩ
R35
15Ω
10Ω
R20
27Ω
SB7
C16
6.4μF
25V
R18
3.3kΩ
C18
150nF
D1
OA91
D2
OA91
C12
6.4μF
25V
R24
39kΩ
C19
150pF
SB3
C20
33nF
R17
68kΩ
R27
220
kΩ
C21
68nF
Record/
playback head
Erase head
C22
68nF
SB8
190/59R

Performance

Overall frequency response

tape speed = $7\frac{1}{2}$in/s	55Hz to 20kHz
tape speed = $3\frac{3}{4}$in/s	55Hz to 12kHz
tape speed = $1\frac{7}{8}$in/s	55Hz to 6kHz

Playback

Output power	4W r.m.s
Sensitivity at base of TR_1, at 1kHz and with tape speed of $7\frac{1}{2}$in/s	400μV
Distortion for 4W output	1·2%
Noise relative to 4W, with volume control set so that a signal produced with 110μA recording current gives 4W output; 3dB bandwidth = 30Hz to 15kHz:	
tape speed = $7\frac{1}{2}$in/s	–60dB
tape speed = $3\frac{3}{4}$in/s	–57dB
tape speed = $1\frac{7}{8}$in/s	–55dB
tape speed = $\frac{15}{16}$in/s	–50dB

Record

Sensitivity at base of TR_1 for 110μA recording current, with volume control at maximum setting	100μV
Distortion at full record level	1%
Recording current level for 4W output, with volume control at maximum setting	40μA

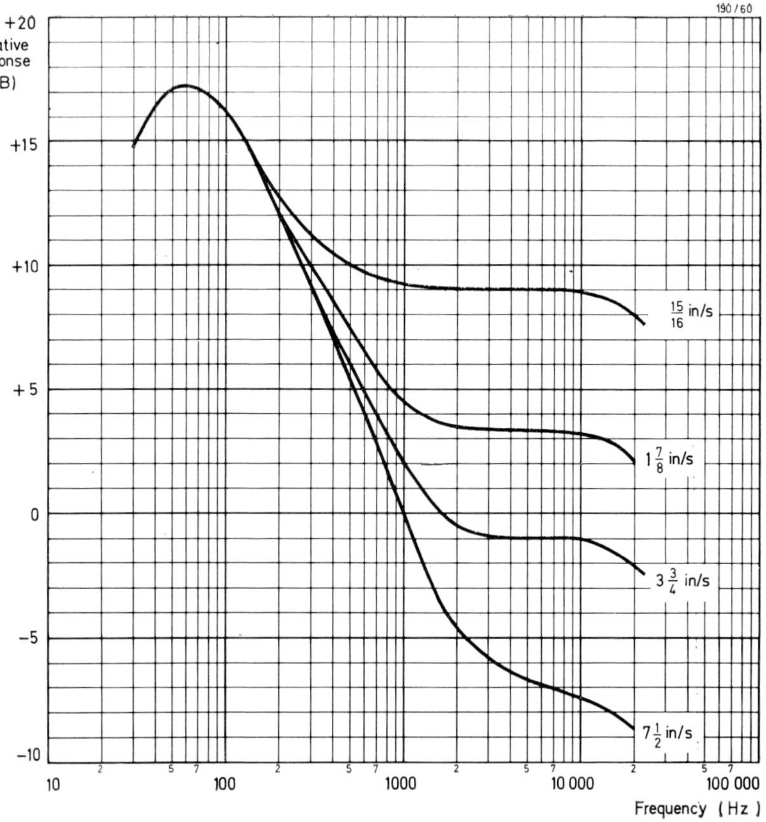

Fig. 60—Playback characteristic

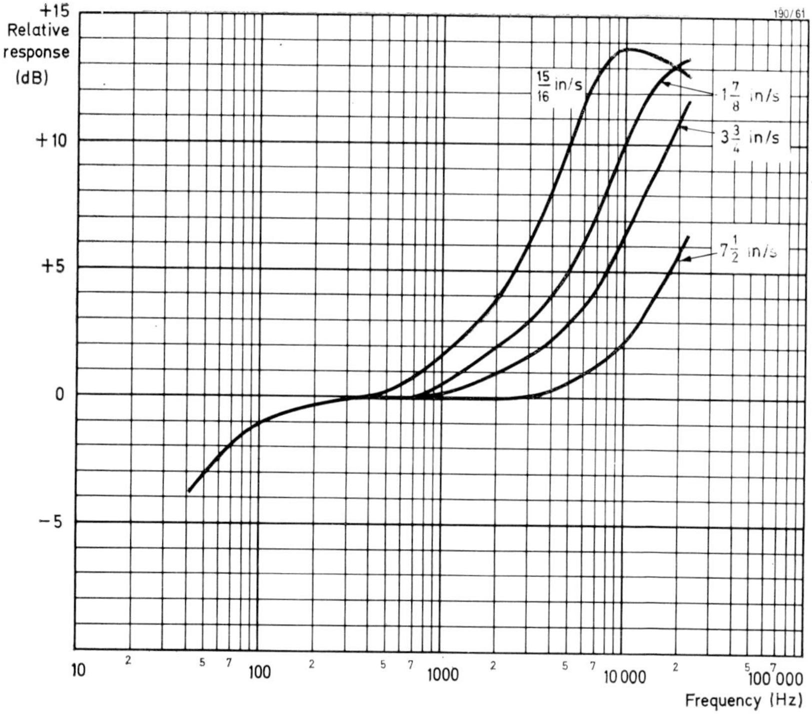

Fig. 61—Record characteristic

TAPE PRE-AMPLIFIER

Circuit Description

Fig. 63 shows a pre-amplifier circuit derived from the tape recorder circuit already described. The oscillator stage in this circuit uses a complementary pair comprising one AC176 and one AC128 transistor.

Performance

The distortion measured at the collector of transistor TR_4 (AC128) in the playback position is shown in Fig. 62. The measurements were made at a speed of $3\frac{3}{4}$in/s, although these characteristics vary little with speed.

At a speed of $7\frac{1}{2}$in/s with the volume control at maximum, the output voltage is 6·55V for an input voltage of 400μV.

The total current drain on playback is 25mA.

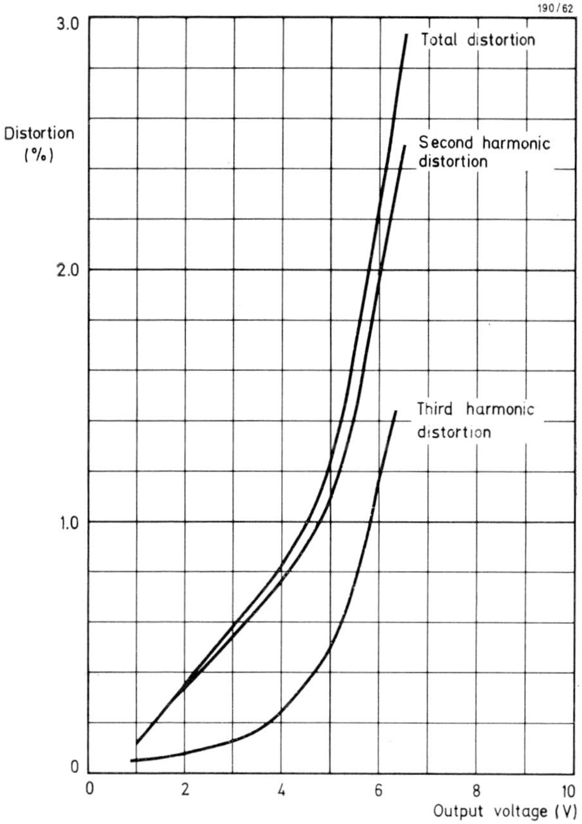

Fig. 62—Variation of harmonic distortion with output voltage for pre-amplifier circuit

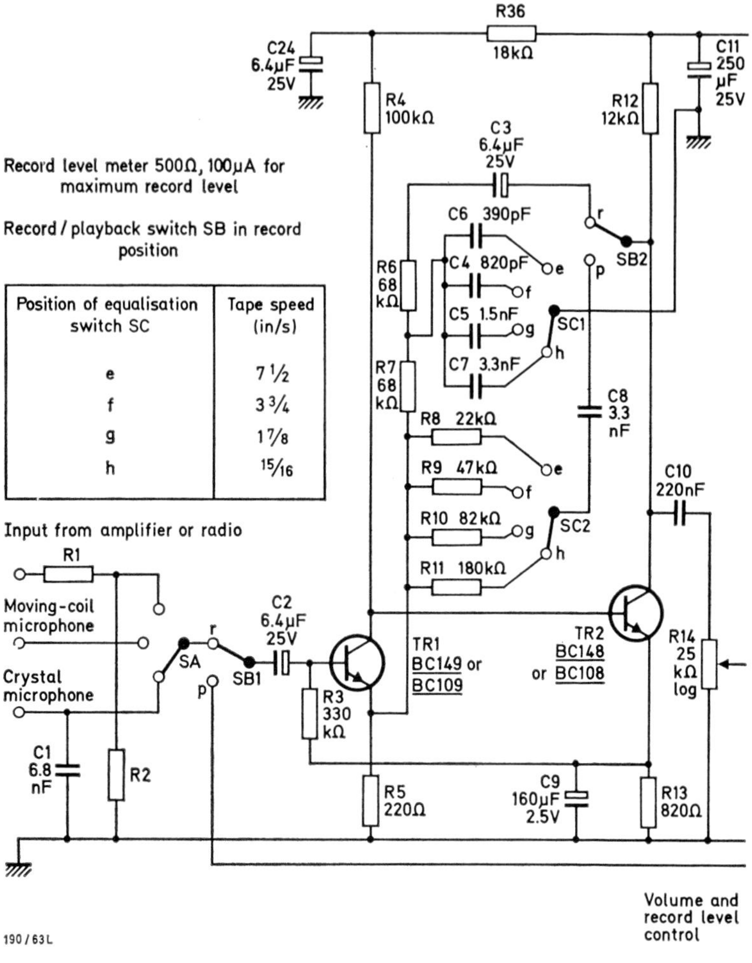

Position of equalisation switch SC	Tape speed (in/s)
e	7 1/2
f	3 3/4
g	1 7/8
h	15/16

Fig. 63—Tape pre-amplifier circuit

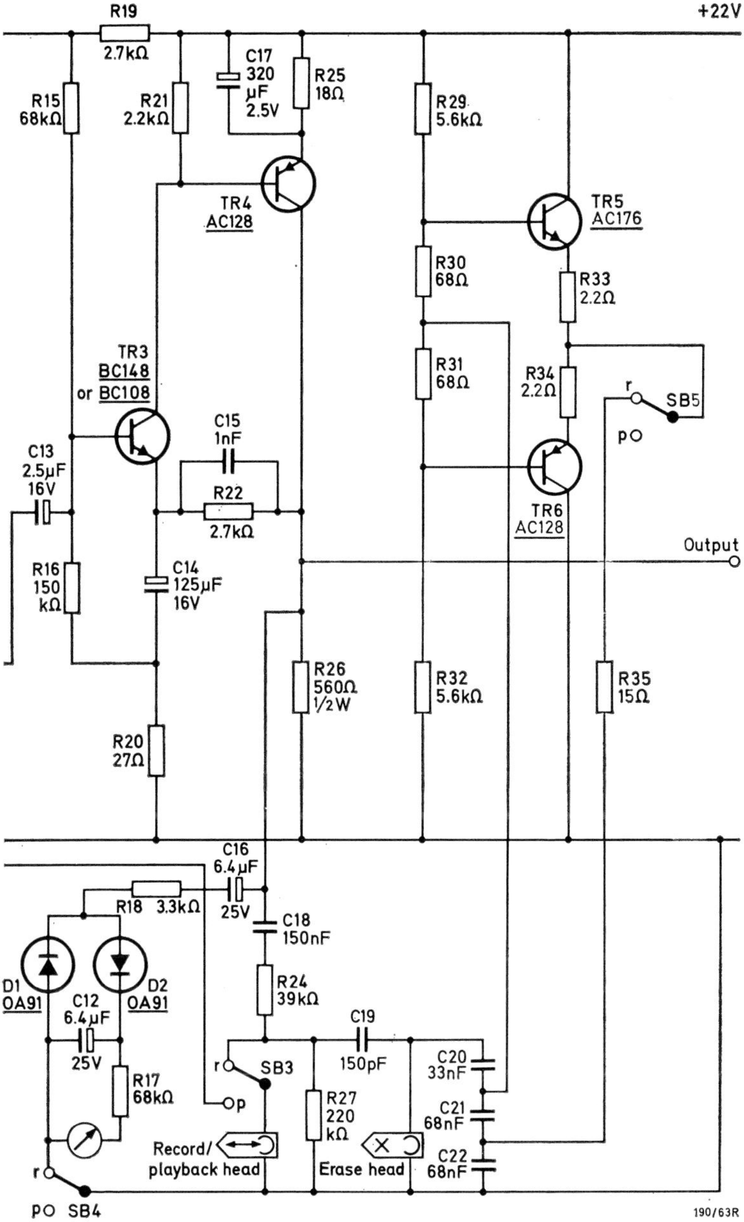
R19
2.7kΩ
+22V
R15
68kΩ
R21
2.2kΩ
C17
320
μF
2.5V
R25
18Ω
R29
5.6kΩ
TR4
AC128
TR5
AC176
R30
68Ω
R33
2.2Ω
TR3
BC148
or BC108
R31
68Ω
R34
2.2Ω
r
SB5
pO
C15
1nF
C13
2.5μF
16V
R22
2.7kΩ
TR6
AC128
Output
R16
150
kΩ
C14
125μF
16V
R26
560Ω
1/2W
R32
5.6kΩ
R35
15Ω
R20
27Ω
C16
6.4μF
25V
R18 3.3kΩ
C18
150nF
D1
OA91
D2
OA91
C12
6.4μF
25V
R24
39kΩ
C19
150pF
C20
33nF
r
SB3
p
R17
68kΩ
R27
220
kΩ
C21
68nF
Record/
playback head
Erase head
C22
68nF
r
pO SB4
190/63R

MODIFICATIONS FOR STEREO OPERATION

In a stereo recording system, the bias and erase frequencies of the two channels must be equal, if the production of audio beat frequencies is to be prevented. Therefore, the bias and erase power for the two channels must be derived from one common oscillator, or from two oscillators which have their frequencies locked together. The former arrangement (Fig. 64) can be used for the pre-amplifier of Fig. 63 only if the AC176/AC128 pair is replaced by an AD161/AD162 pair and the emitter resistors reduced accordingly. In the latter arrangement (Fig. 65), two AC176/AC128 pairs will operate satisfactorily.

Bias and erase power from a common oscillator

A circuit that uses a common oscillator for stereo operation is given in Fig. 64. This shows that the two tuned erase heads, which form the load and feedback circuits of one of the complementary pairs of transistors, are in parallel. The bias for the two recording heads, however, is still derived from the erase voltages across the erase heads as in the mono design. This has the advantage that the other pair of complementary transistors used for the output stage during playback may now be used as a monitoring facility.

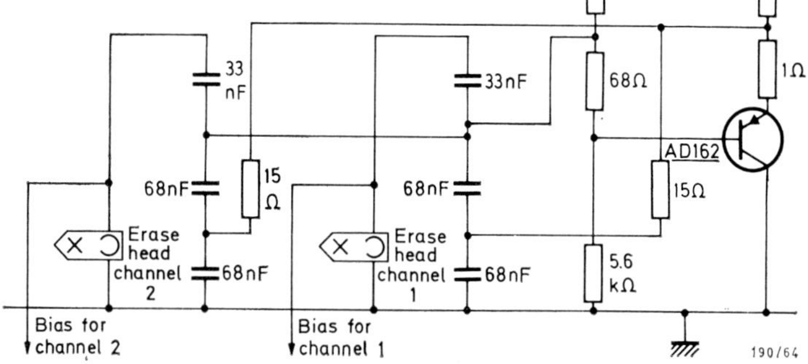

Fig. 64—Bias and erase power from common oscillator for stereo operation

Bias and erase power from two oscillators with locked frequencies

A circuit that uses two locked identical oscillators for stereo operation is given in Fig. 65. Locking of the oscillators is achieved by connecting together the two feedback paths.

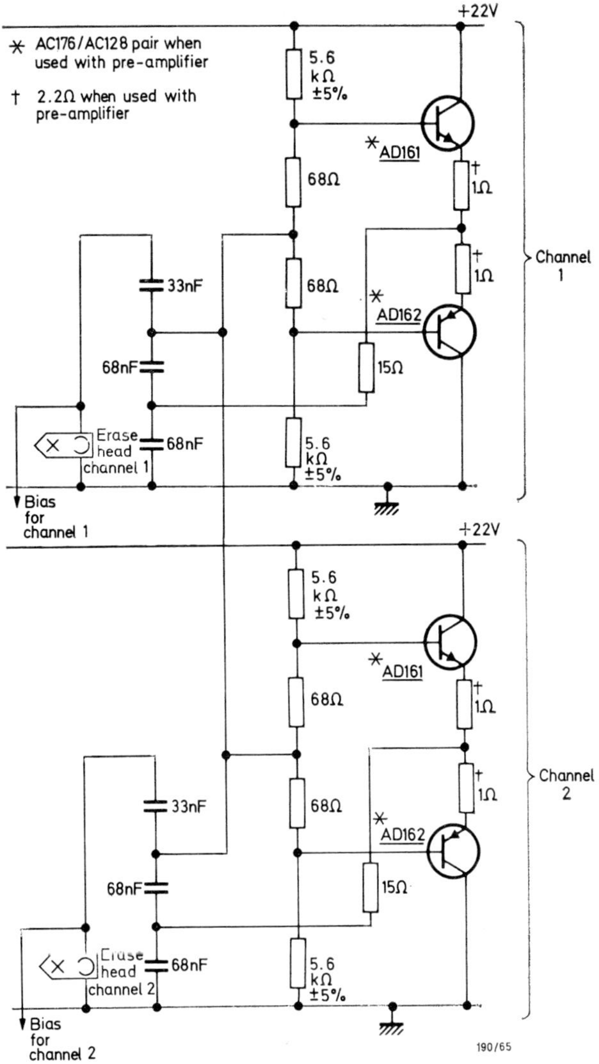

Fig. 65—Bias and erase power from two oscillators with locked frequencies for stereo operation

Recording level indicator

Fig. 66 shows a recording level indicator for stereo operation. The meter gives a reading proportional to the peak-to-peak voltage at the collector of the driver transistor in the channel which has the highest recording level. For an input of 4V r.m.s. (this corresponds to the r.m.s. voltage at the collector of the driver transistor when distortion due to tape saturation becomes significant) the d.c. current through the meter is 95μA.

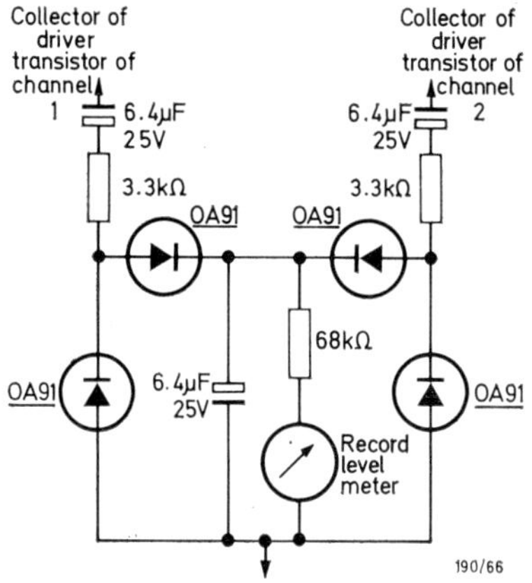

Fig. 66—Recording level indicator for stereo operation

A.G.C. CIRCUIT FOR TAPE RECORDERS

When recording on tape, the current in the recording head must not exceed the value at which tape saturation occurs if severe distortion is to be avoided. This can be prevented by monitoring the current by means of a meter or magic eye, and manually adjusting the recording level. Another method is to use a circuit that electronically reduces the gain of the amplifier when the optimum recording level is exceeded.

To ensure that the distortion is recorded for only a short time, the gain control must respond very rapidly to the overload signal. When the gain has been reduced, however, the recovery time must be long, for otherwise volume compression will result, and loud and soft passages of music will be recorded with equal loudness.

A gain control circuit that uses a minimum of components and can produce an attenuation of 40dB within 150 milliseconds is shown in Fig. 67. The recovery rate does not exceed 6dB per minute. With a faster recovery rate, the same circuit arrangement is suitable as a volume compressor in other applications.

Circuit Description

Gain control is achieved by means of an attenuator containing a diode. Controlling the direct current through the diode controls its slope resistance, and, consequently, the amount of attenuation.

The severe harmonic distortion caused by the exponential characteristic of one diode can be reduced by using two diodes in a reverse-parallel arrangement. In the circuit in Fig. 67, the base-emitter junction of a transistor is used as one of the diodes. Because the transistor is needed as a d.c. amplifier, this arrangement saves the expense of one diode.

Part of the amplifier output is rectified and used to provide the gain control signal. The time constants of the rectifier circuit are such as to give the high rate of attenuation and the long recovery time.

In Fig. 67 unit A is the audio amplifier and R_2 represents its input resistance, which should be about 7kΩ.

The variable attenuator comprises R_1, C_1, D_1 and part of the transistor TR_1. Because C_3 is a short-circuit to a.c. signals, the base-emitter junction of the transistor is, in effect, in reverse-parallel with D_1 for all a.c. signals.

Resistors R_5 and R_6 form a fixed attenuator by means of which a constant fraction of the output is fed back to the control circuit. The feedback voltage is rectified by D_2, charges C_4, and is applied through R_3 to the base of TR_1. Hence, the voltage across C_4 controls the current through TR_1 and D_1.

With an input signal well below that which causes tape saturation, the voltage across C_4 is low. Therefore, the transistor current is also low and

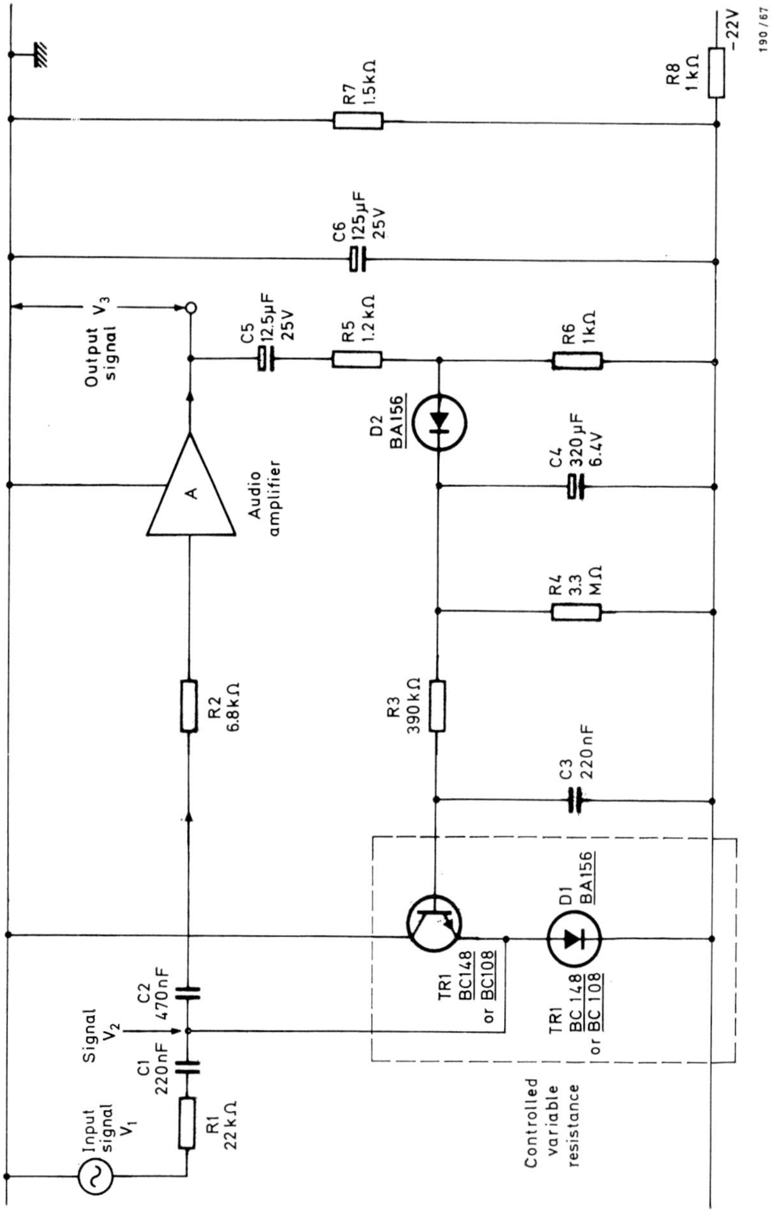

Fig. 67—Automatic gain control circuit

the slope resistances of it and the diode are high. Consequently, the shunting effect that the diode and transistor have on the input to the amplifier is negligible.

As the input signal is increased, the voltage across C_4 and the current through TR_1, will also increase. Eventually, the transistor current will reach a value which produces a significant reduction in the slope resistance of D_1. The voltage drop across R_1 will be increased and there will be an appreciable attenuation of the signal to the amplifier. This occurs when the amplifier has an output of approximately 2V r.m.s.

A further increase in the input produces another rise in the voltage across C_4. The consequent increase in transistor current now produces a rapid reduction in the diode slope resistance. The shunting effect of the transistor and diode on the input to the amplifier is greatly increased, and V_2 is much less than V_1.

Once attenuation has started, voltages V_2 and V_3 remain practically constant. Over the full range of attenuation, 40dB, they increase slightly because of slight increases in V_{BE}, the forward voltage of D_1, at higher currents (of the order of 300μA), and the base current through R_3.

The attack time is largely determined by the time-constant of C_4 and the source impedance of the charging circuit, which is R_5 in parallel with R_6. For an attack time of about 200ms, the source impedance should be approximately 600Ω.

The recovery time is determined by the time-constant of C_4 and its discharge circuit—R_4 in parallel with R_3 and the slope resistances of TR_1 and D_1. Over a wide range of the attenuation, R_3 is much larger than the slope resistances of the diode and transistor. The rate of discharge of C_4 is then controlled by the value of R_3, and, consequently, the rate of change of attenuation is constant. At very low transistor currents, the slope resistances of the diode and the transistor greatly exceed R_3, and the rate of discharge of C_4 via the transistor is very low. This tends to make the rate of change of attenuation also very low. Under these conditions, however, R_4 provides an alternative discharge path. Hence, the rate of change of attenuation has a low limit that is determined by the value of R_4. When R_3 is large with respect to the slope resistances of the diode and the transistor, the effect of R_4 is negligible.

The transistor is a BC148 or BC108 n-p-n planar type, which, with its low leakage current and high gain, makes this simple arrangement possible. The silicon diodes also are used because of their low leakage currents.

In Fig. 67, the gain is controlled by attenuating the signal before it reaches the audio amplifier. This has the disadvantage that, when the signal is attenuated, the noise originating in the first transistor of the amplifier is unaffected. A better arrangement would be to have the attenuation after the first stage of the amplifier, thus reducing noise and signal simultaneously. This arrangement is shown in Fig. 68. A high voltage is required to operate the attenuating circuit; this will probably be available only at the output of the stage feeding the recording head.

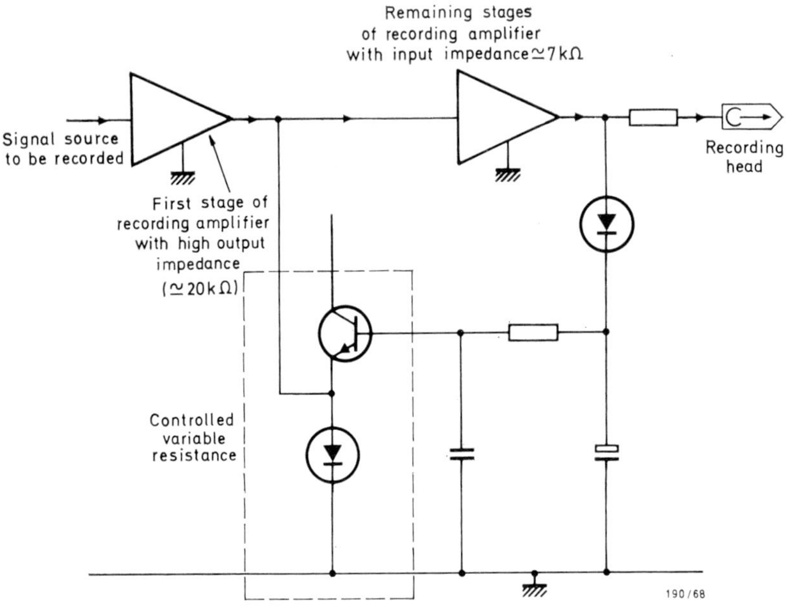

Fig. 68—Incorporation of gain control circuit in a tape amplifier

Use of Electrolytic Capacitors in Timing Circuits

The maximum leakage currents quoted in published data on electrolytic capacitors are too high for use in long time-constant circuits. Experience suggests, however, that the leakage currents of capacitors taken off the shelf will be low enough or will soon fall to a suitably low value. Therefore, electrolytic capacitors of a suitable type are readily available. At the worst, a tape recorder which has been stored for a very long time will have an initial recovery time slightly shorter than normal, but this will rapidly approach the normal value.

Performance

Where the circuit controls the gain, the signal V_2 is not maintained exactly constant nor completely independent of the amount of attenuation.

This is shown in Fig. 69 by the input/output characteristics of the circuit with high-gain and low-gain transistors. When a low-gain transistor is used, an attenuation of 40dB is accompanied by an increase of 7dB in the output; with a high-gain transistor, the corresponding increase in output is 4·5dB.

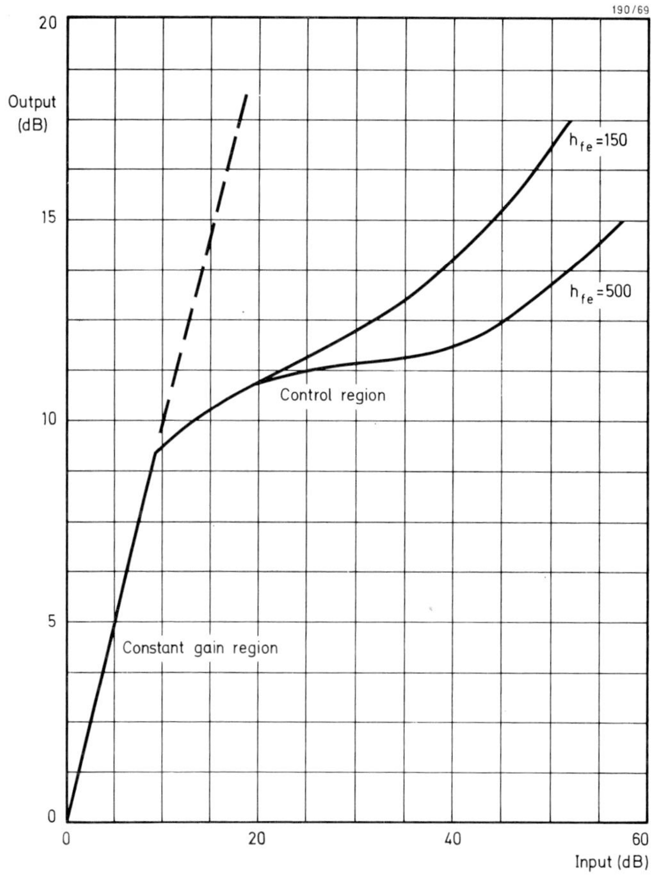

Fig. 69—Variation of output with input for high-gain and low-gain transistors

The change of attenuation with time is shown in Fig. 70. This diagram shows that the rate of change of attenuation is substantially constant over much of the range. It does not exceed 6dB per minute and is almost independent of the gain of the transistor.

The variation of total harmonic distortion of V_2 with the amount of attenuation for various values of V_2 is given in Fig. 71. These curves show that V_2 should not exceed 4mV r.m.s. if the total harmonic distortion is not to exceed 2%.

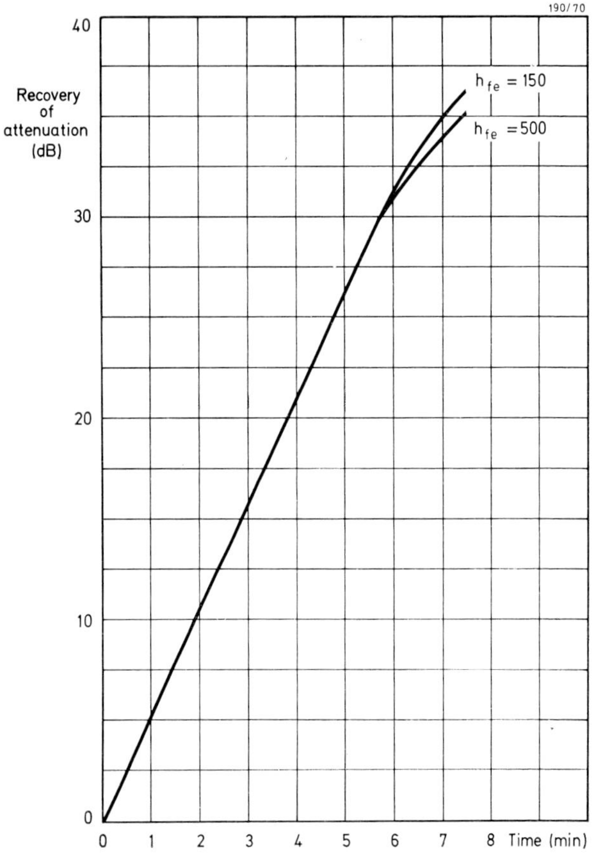

Fig. 70—Decrease of attenuation with time for high-gain and low-gain transistors

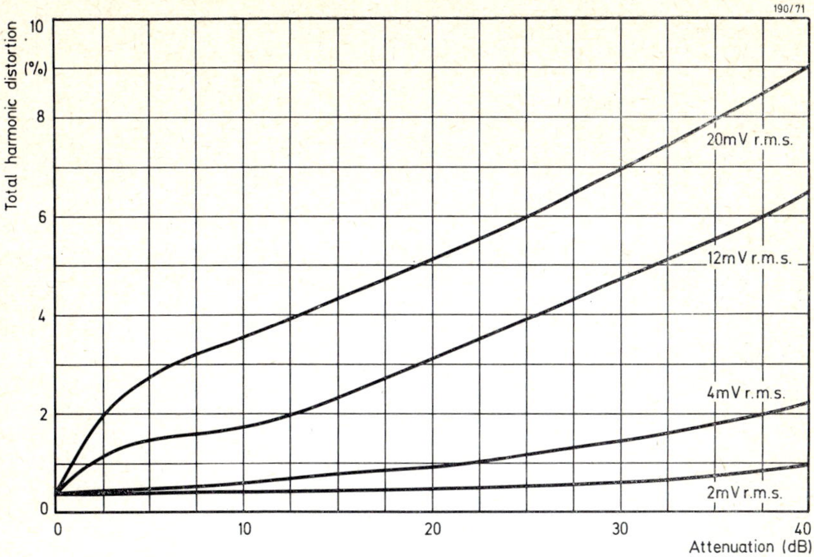

Fig. 71—Variation of total harmonic distortion of V_2 with amount of attenuation, for various values of V_2

COMPONENT LISTS FOR CIRCUITS IN CHAPTER 5

4W Tape Recorder (pages 90 and 91)

Heads

Record/playback head	Marriot X/RPS/36
Erase head	Marriot X/ES/11

Transistors and Diodes

Circuit reference	Mullard type
TR_1	BC149 or BC109
TR_2	BC148 or BC108
TR_3	BC148 or BC108
TR_4	AC128
TR_5	AD161
TR_6	AD162
D_1	OA91
D_2	OA91
Full-wave bridge rectifier	BY164

Resistors

Mullard CR25 series (0·3W) ±5% except R_{26} which is CR37 series (0·5W) ±5%

Circuit reference	Value	Circuit reference	Value
R_1 } to suit input,	500kΩ	R_{20}	27Ω
R_2 } typically	22kΩ	R_{21}	2·2kΩ
R_3	330kΩ	R_{22}	2·7kΩ
R_4	100kΩ	R_{23} preset potentiometer	50Ω
R_5	220Ω		
R_6	68kΩ	R_{24}	39kΩ
R_7	68kΩ	R_{25}	18Ω
R_8	22kΩ	R_{26}	560Ω
R_9	47kΩ	R_{27}	220kΩ
R_{10}	82kΩ	R_{28} thermistor type VA1077	32Ω
R_{11}	180kΩ		
R_{12}	12kΩ	R_{29}	5·6kΩ
R_{13}	820Ω	R_{30}	68Ω
R_{14} logarithmic potentiometer	25kΩ	R_{31}	68Ω
		R_{32}	5·6kΩ
R_{15}	68kΩ	R_{33}	1Ω
R_{16}	150kΩ	R_{34}	1Ω
R_{17}	68kΩ	R_{35}	15Ω
R_{18}	3·3kΩ	R_{36}	18kΩ
R_{19}	2·7kΩ		

Capacitors

Circuit reference	Value	Description	Mullard type
C_1	6·8nF	polyester	C296AC/A6K8
C_2	6·4μF	electrolytic 25V	C426AR/F6·4
C_3	6·4μF	electrolytic 25V	C426AR/F6·4
C_4	820pF	polystyrene	C295AC/820E
C_5	1·5nF	polyester	C296AC/A1K5
C_6	390pF	polystyrene	
C_7	3·3nF	polyester	C296AC/A3K3
C_8	3·3nF	polyester	C296AC/A3K3
C_9	160μF	electrolytic 2·5V	C426AR/A160
C_{10}	220nF	polyester	C296AA/A220K
C_{11}	250μF	electrolytic 25V	C437AR/F250
C_{12}	6·4μF	electrolytic 25V	C426AR/F6·4
C_{13}	2·5μF	electrolytic 16V	C426AR/H2·5
C_{14}	125μF	electrolytic 16V	C426AR/E125
C_{15}	1nF	polyester	C296AC/A1K
C_{16}	6·4μF	electrolytic 25V	C426AR/F6·4
C_{17}	320μF	electrolytic 2·5V	C426AR/A320
C_{18}	150nF	polyester	C296AA/A150K
C_{19}	150pF	ceramic plate	C333CH C150E
C_{20}	33nF	polyester	C296AA/A33K
C_{21}	68nF	polyester	C296AA/A68K
C_{22}	68nF	polyester	C296AA/A68K
C_{23}	640μF	electrolytic 16V	C437AR/E640
C_{24}	6·4μF	electrolytic 25V	C426AR/F6·4
C_{25}	640μF	electrolytic 25V	C437AR/F640

Mains Transformer

The transformer used in this circuit may be obtained from the following manufacturers under the type numbers given:

Colne Electric Limited	20042
Drake Transformers Limited	351-245
Gardners Transformers Limited	GR 97182
Parmeko Limited	P 3203

Heatsinks

Each output transistor should be mounted on a 1·5mm (16 gauge) bright aluminium heatsink with an area of 26cm^2. The driver transistor TR_4 should be mounted by means of a heat clip onto a 1·5mm bright aluminium heatsink with an area of 13cm^2.

Tape Pre-amplifier (pages 96 and 97)

As for the tape recorder amplifier (page 108) except:

Transistors

TR_5	AC176
TR_6	AC128

Resistors

R_{23}	omitted
R_{28}	omitted
R_{33}	2·2Ω
R_{34}	2·2Ω

Capacitors

C_{23}	omitted

A.G.C. Circuit (page 102)

Transistor and Diodes

Circuit reference	Mullard type
TR_1	BC148 or BC108
D_1	BA156
D_2	BA156

Resistors

Mullard CR25 series (0·3W) ±5%

Circuit reference	Value
R_1	22kΩ
R_2	see text
R_3	390kΩ
R_4	3·3MΩ (±10%)
R_5	1·2kΩ
R_6	1kΩ
R_7	1·5kΩ
R_8	1kΩ

Capacitors

Circuit reference	Value	Description	Mullard type
C_1	220nF	polyester	C296AA/A220K
C_2	470nF	polyester	C296AA/A470K
C_3	220nF	polyester	C296AA/A220K
C_4	320μF	electrolytic 6·4V	C426AR/C320
C_5	12·5μF	electrolytic 25V	C426AR/F12·5
C_6	125μF	electrolytic 25V	C426AR/E125

CHAPTER 6

A.F. AMPLIFIERS FOR CAR RADIOS

The chapter describes two a.f. amplifiers, a 5W class A amplifier and a 6W class B amplifier, suitable for use in a car radio.

Component lists for the circuits are given at the end of the chapter.

5W CLASS A AMPLIFIER

The circuit of a 5W class A audio amplifier for use in a car radio is given in Fig. 72. It contains two BC148 or BC108 transistors and one BD121 transistor, and is designed to operate from a nominal 14V supply.

Circuit Description

Direct coupling is used between stages, and the base-emitter voltage of TR_1 serves as a reference for the emitter voltage of TR_3. Consequently, the emitter voltage of TR_1 controls the direct current through TR_3 and determines its operating point.

Variable resistor R_{11} is used to set the quiescent current—900mA— of the output stage. The loudspeaker is driven by means of a centre-tapped choke which incorporates a small winding to provide feedback to the first

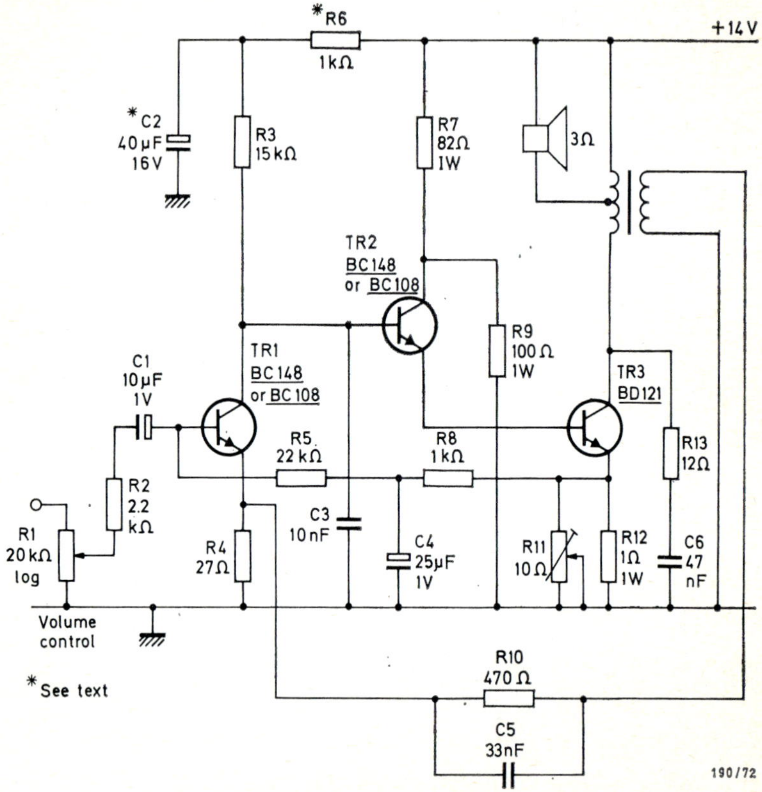

Fig. 72—5W class A audio amplifier circuit for use in a car radio

stage. The choke has an inductance of 30mH and a d.c. resistance of 1Ω. The voltage ratio of choke to secondary is 13:1. Resistor R_{10} and capacitor C_5 limit the frequency response of the feedback loop and amplifier.

Resistor R_{13} and capacitor C_6 are included to ensure that the load remains resistive, and hence the circuit stable, at high frequencies. As low-frequency stability depends upon the source impedance, a 2·2kΩ resistor, R_2, is connected in series with the wiper of the volume control R_1.

Decoupling between resistors R_3 and R_7 is effected by resistor R_6 and capacitor C_2. The values given for R_6 and C_2 in the circuit diagram may have to be modified as they depend upon the decoupling at the supply input. The voltage drop across R_6 should not exceed 1V; if it does, the value of R_3 must be reduced.

Performance

Output power with symmetrical clipping	5W r.m.s.
Sensitivity (input for 5W output with volume control set for maximum volume)	43mV
Frequency response (—3dB points)	50Hz to 10kHz
Input resistance (including 2·2kΩ resistor but excluding volume control)	13kΩ
Negative feedback	12dB
Distortion at 5W output (see Fig. 73)	2·5%

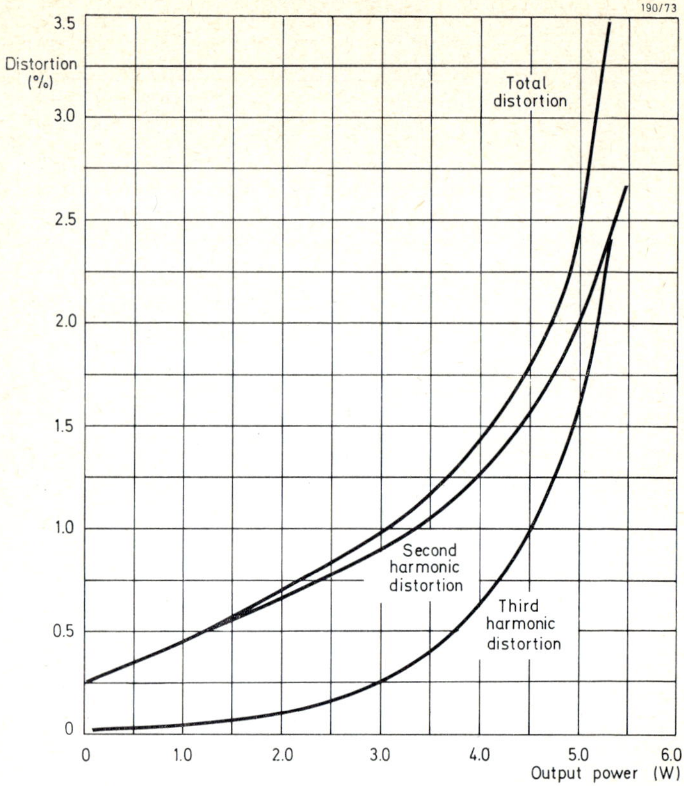

Fig. 73—Variation of harmonic distortion with output power

6W CLASS B AMPLIFIER

The circuit of a 6W class B audio amplifier for use in a car radio is shown in Fig. 74. The amplifier contains an AD161/AD162 complementary output pair, a BFX88 driver transistor, and a BC148 or BC108 transistor as the first-stage amplifier.

Circuit Description

The amplifier is a conventional four-transistor circuit with one exception: the decoupling capacitor C_2 in the input stage is returned to the emitter of transistor TR_1 instead of to chassis. The purpose of this is two-fold: firstly, to neutralise the effect of ripple on the h.t. line, and secondly, to bootstrap the base resistor, thereby increasing the input resistance.

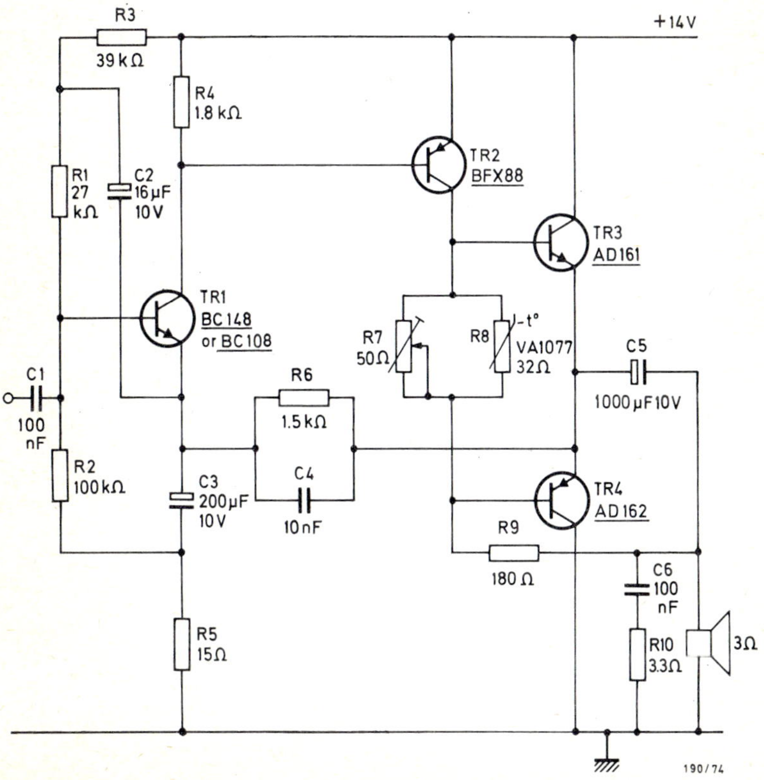

Fig. 74—6W class B audio amplifier circuit for use in a car radio

Performance

Output power	6W r.m.s.
Sensitivity for full output	52mV
Frequency response at 3W (—3dB points)	74Hz to 10·6kHz
Input resistance at 1kHz	100kΩ
Feedback	20dB
Distortion at 6W output (see Fig. 75)	1·2%
Quiescent current in output stage only	30mA
Total current at 6W output	700mA

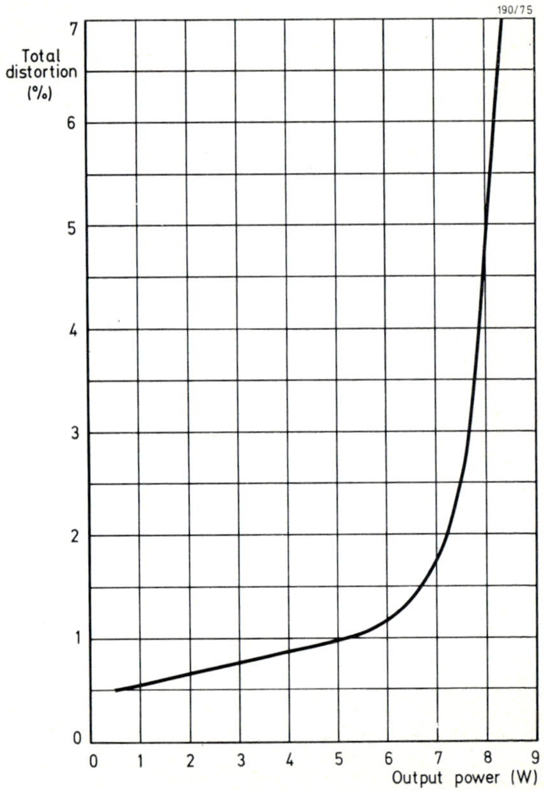

Fig. 75—Variation of total harmonic distortion with output power

COMPONENT LISTS FOR CIRCUITS IN CHAPTER 6

5W Class A Amplifier (page 112)

Transistors

Circuit reference	Mullard type
TR_1	BC148 or BC108
TR_2	BC148 or BC108
TR_3	BD121

Resistors

Mullard CR25 series (0·3W) ±5% except R_7, R_9, and R_{12} which are 1W components

Circuit reference	Value	Circuit reference	Value
R_1 logarithmic potentiometer	20kΩ	R_8	1kΩ
		R_9	100Ω
R_2	2·2kΩ	R_{10}	470Ω
R_3	15kΩ	R_{11} preset potentiometer	10Ω
R_4	27Ω		
R_5	22kΩ	R_{12}	1Ω
R_6	1kΩ	R_{13}	12Ω
R_7	82Ω		

Capacitors

Circuit reference	Value	Description	Mullard type
C_1	10μF	electrolytic 1V	C426AR/E10
C_2	40μF	electrolytic 16V	C426AR/E40
C_3	10nF	polyester	C296AA/A10K
C_4	25μF	electrolytic 1V	C426AR/C25
C_5	33nF	polyester	C296AA/A33K
C_6	47nF	polyester	C296AA/A47K

Heatsinks

The values of R_7 and R_9 ensure that the maximum permissible junction temperature of transistor TR_2 is not exceeded and so no heat clip is required. The output transistor, however, has a maximum dissipation of 11W; therefore at an ambient temperature of 60°C, a 1·5mm (16 gauge) bright aluminium heatsink with an area of $50cm^2$ is required.

6W Class B Amplifier (page 115)

Transistors

Circuit reference	Mullard type
TR_1	BC148 or BC108
TR_2	BFX88
TR_3	AD161
TR_4	AD162

Resistors

Mullard CR25 series (0·3W) ±5% except R_9 which is CR37 series (0·5W) ±5%

Circuit reference	Value	Circuit reference	Value
R_1	27kΩ	R_7 preset potentiometer	50Ω
R_2	100kΩ		
R_3	39kΩ	R_8 thermistor type VA1077	32Ω
R_4	1·8kΩ		
R_5	15Ω	R_9	180Ω
R_6	1·5kΩ	R_{10}	3·3Ω

Capacitors

Circuit reference	Value	Description	Mullard type
C_1	100nF	polyester	C296AA/A100K
C_2	16μF	electrolytic 10V	C426AR/D16
C_3	200μF	electrolytic 10V	C426AR/D200
C_4	10nF	polyester	C296AA/A10K
C_5	1000μF	electrolytic 10V	C437AR/D1000
C_6	100nF	polyester	C296AA/A100K

Heatsinks

The driver dissipation is only 350mW, so no heatsink is required for this transistor. If mica washers are used, each output transistor should be mounted on a 1·5mm (16 gauge) bright aluminium heatsink with an area of $16cm^2$.

CHAPTER 7

HIGH-QUALITY AUDIO EQUIPMENT

In this chapter audio amplifiers with output powers of 10W, 15/20W, 25W, 35W, and 50W are described, together with two pre-amplifiers suitable for use with these amplifiers. The chapter includes auxiliary audio amplifiers, control circuits and filters. Short-circuit protection of high-power amplifiers is also described.

In the construction of high-quality equipment attention must be paid to component layout; the recommendations made in this chapter should be carefully followed.

Components lists for the circuits are given at the end of the chapter.

WHAT IS HI-FI?

'Hi-Fi' is a term much used in conjunction with sound-reproducing equipment, and greatly beloved of advertisers. It is an undefined term and has been frequently abused by applying it to inferior equipment. For this reason 'high quality' has been used in this book—an equally undefined but less abused term.

'Hi-Fi' is a contraction of high fidelity which was a term applied to sound-reproducing equipment that gave a realistic recreation of the original sound. Thus to some extent 'Hi-Fi' is what you make of it. If a particular system makes to your ears a life-like sound, then you could justifiably call it 'Hi-Fi'. The problem is that the system may not sound life-like to a more critical listener.

Because of this subjective element in determining just what is 'Hi-Fi' (or high quality), attempts have been made to define minimum performance standards for equipment. At present the best-known standard in Europe is that issued by the Deutscher Industrie Normenausschuss, DIN 45-500. This standard lays down some minimum performance figures that should be met by equipment claiming to be 'high quality'. For example:

the frequency response of the system should be wide (40Hz to 16kHz, $\pm 1\cdot5$dB, for power amplifiers);

low noise (50dB below 100mW for amplifiers of less than 20W output);

low distortion (not greater than 1% for power amplifiers from full power down to −20dB over a frequency range 40Hz to 12·5kHz);

and so on. Recommendations are also made about such matters as channel balance limits in stereo systems, minimum useful output powers, directional characteristics of microphones, mechanical speed stability ('wow' and 'flutter') of turntables and tape decks, and tolerances for equalised inputs for disc pick-ups.

The DIN standard has been criticised by many specialists as being too lenient for true high-quality sound reproduction. Nevertheless, the fact remains that a considerable amount of so-called 'Hi-Fi' equipment fails to meet its requirements in one or more respects.

It is important that the parts of an audio system should be of equal quality. The effects of a high-quality amplifier will be lost if inferior speakers are used. Again, an extended bass response from the amplifier and speakers may reveal 'rumble' from a poor turntable. Also, to some extent the law of diminishing returns applies to audio equipment. After certain minimum requirements have been met, improvements to performance cost more and more money. If you can hear the difference in sound quality between cheaper and more expensive equipment, and can afford it, then it is better to buy the more expensive equipment. If not, an extra kilohertz on the treble response should not influence you—your ears should be the judge. And remember, the better your equipment, the cleaner you have to keep your records.

The circuits in this chapter have been designed in accordance with current good audio-engineering practice. At the same time, unnecessary elaboration has been avoided. The circuits therefore represent a sensible compromise between price and performance.

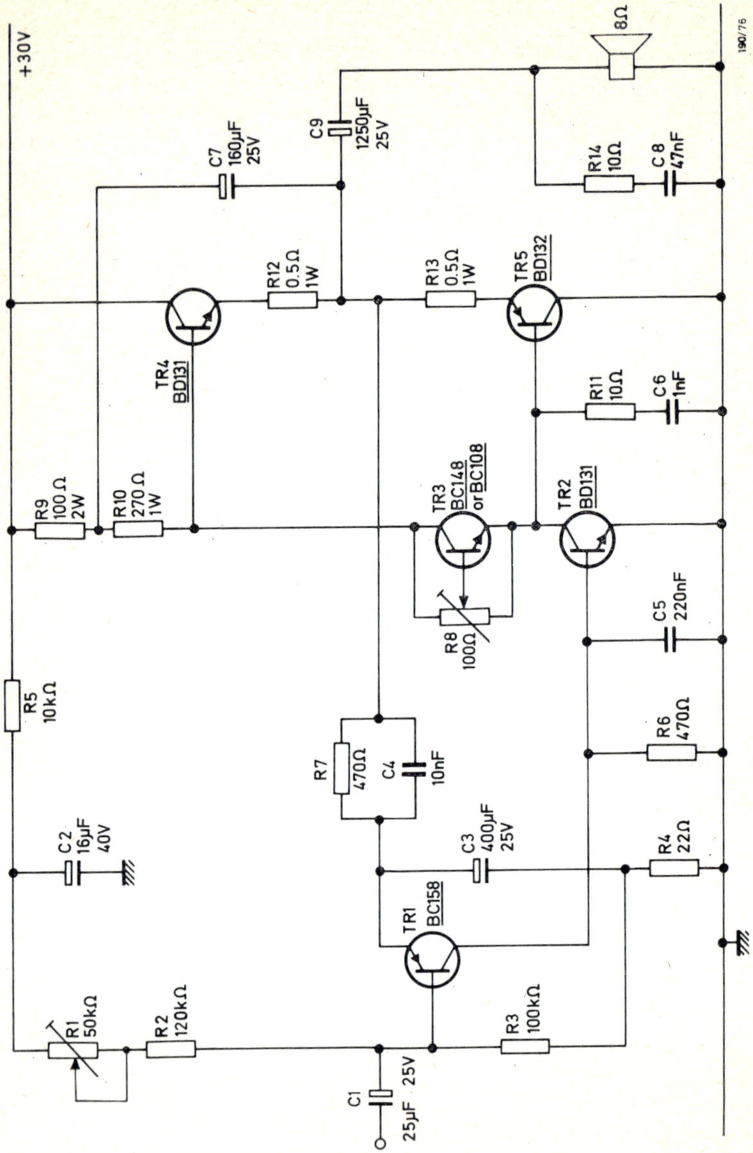

Fig. 76—10W high-quality audio amplifier circuit

10W AUDIO AMPLIFIER

The circuit for a 10W high-quality audio amplifier is shown in Fig. 76. The use of a complementary pair of output transistors gives a much simpler circuit than the quasi-complementary circuit usually employed in class B high-quality amplifiers. In addition, the greater symmetry of the circuit gives lower distortion, particularly in the crossover region.

Circuit Description

The circuit uses a BD131/BD132 complementary pair to deliver an output power of 10W at a distortion level of less than 0·1%. The familiar four-transistor, directly-coupled configuration is used, but an additional transistor TR_3 is included to stabilise the quiescent current of the output transistors. Preset potentiometer R_1 sets the mid-point voltage for symmetrical clipping.

Power Supply

Fig. 77 shows a suitable power supply. The supply requirements are:

Total quiescent current for one channel	55mA
Full load current for one channel (at 30V)	600mA
Supply voltage	30V
Maximum supply voltage	36V

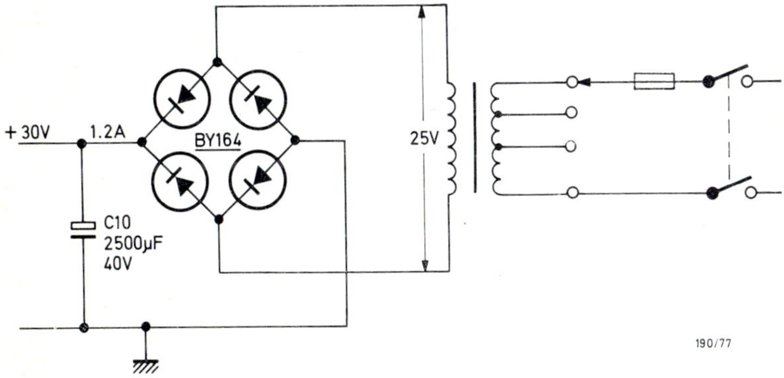

Fig. 77—Power supply circuit for 10W amplifier

Performance

Output power (with 30V supply)	10W r.m.s.
Sensitivity for full output	430mV
Frequency response at 2W output (–3dB points)	20Hz to 35kHz
Input resistance	90kΩ
Distortion at 10W output (see Fig. 78)	<0·1%
Quiescent current of output transistors	15mA

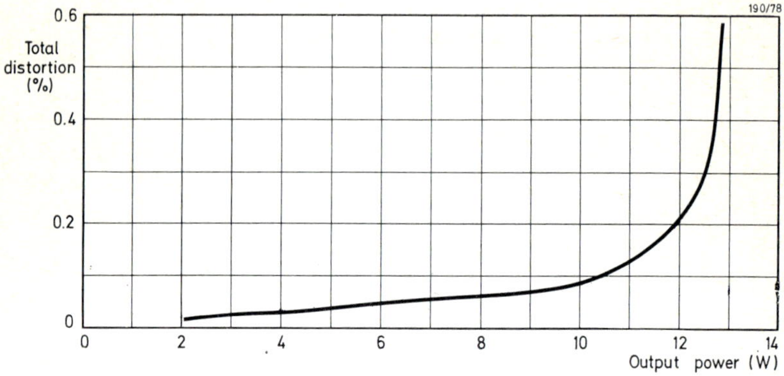

Fig. 78—Variation of total harmonic distortion with output power

15/20W AUDIO AMPLIFIER

A high-quality 15/20W audio amplifier circuit is shown in Fig. 79. This amplifier is designed to operate in class A into an 8Ω load, giving an output of 15W. With a 4Ω load, however, the circuit will operate in class AB to give 20W output power. The total harmonic distortion is less than 0·1% at full output.

The amplifier will withstand normal overdrive conditions and does not require additional protection against short-circuit conditions.

Circuit Description

The output transistors, a matched pair of BD181 devices, are driven by two BC338 transistors. These transistors are preceded by a phase-splitting stage using two BC147 transistors, and a pre-amplifier stage, a BC158.

A BC158 transistor is used in the pre-amplifier stage. A.C. and d.c. feedback is applied, giving an input impedance of 150kΩ. This stage also provides stabilisation of the mid-point voltage, which is set by adjustment of the preset resistor R_1.

Phase splitting is achieved by transistors TR_2 and TR_3 connected in a long-tail-pair configuration, and local feedback is applied by resistor R_{11}. A supply voltage higher than that applied to the output stage is necessary to increase the voltage swing available.

The driver transistors TR_4 and TR_5 are bootstrapped by capacitors C_{11} and C_{12} to reduce the dissipation, especially under overdrive and short-circuit conditions. The output transistors TR_6 and TR_7 are matched BD181 transistors designed for high power dissipation at high supply voltage. Resistors R_{20} and R_{21} ensure that these transistors can operate up to their V_{CER} rating.

Performance

Input impedance	150kΩ
Input sensitivity	
for 15W output into an 8Ω load	360mV
for 20W output into a 4Ω load	295mV
Total harmonic distortion	see Fig. 80
Frequency response	
−1dB points at $\frac{1}{4}\times$ full-power output	40Hz to 16kHz
−3dB points at $\frac{1}{4}\times$ full-power output	20Hz to 30kHz
Signal-to-noise ratio, at 50mW output power and source impedance 1kΩ	>80dB

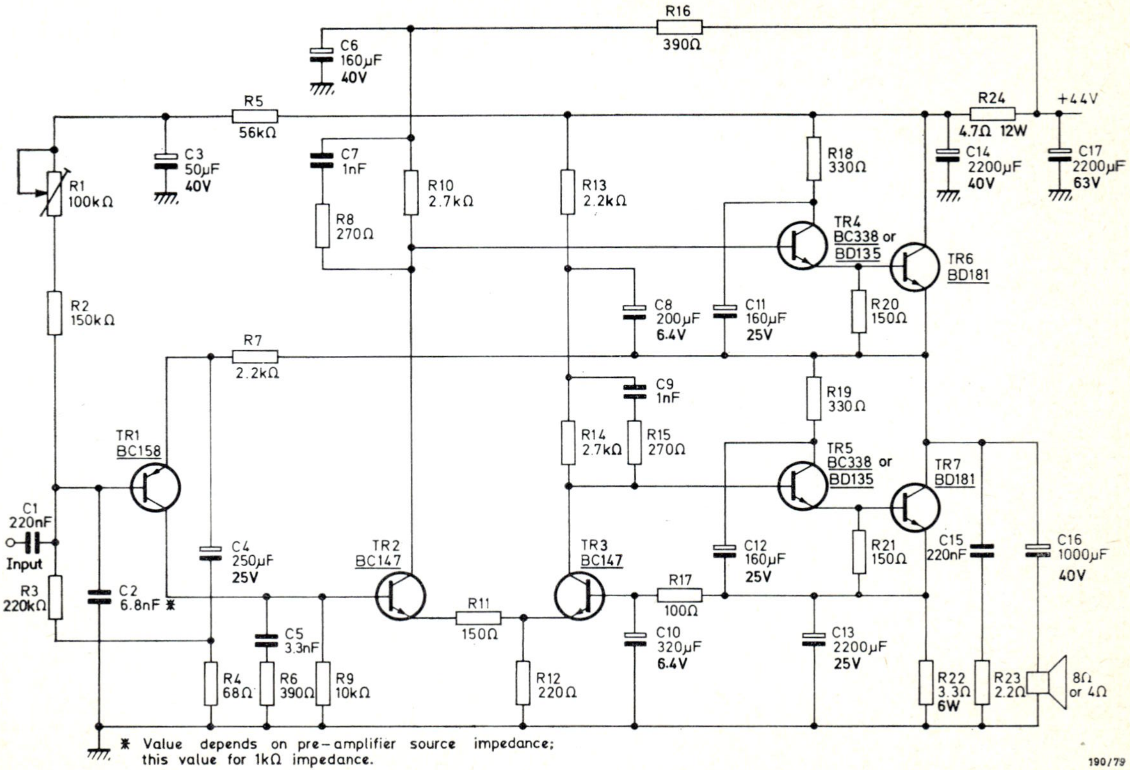

Fig. 79—15/20W high-quality audio amplifier circuit

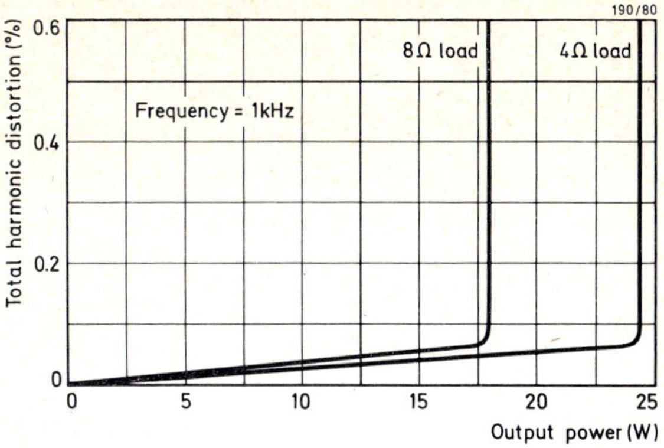

Fig. 80—Variation of total harmonic distortion with output power

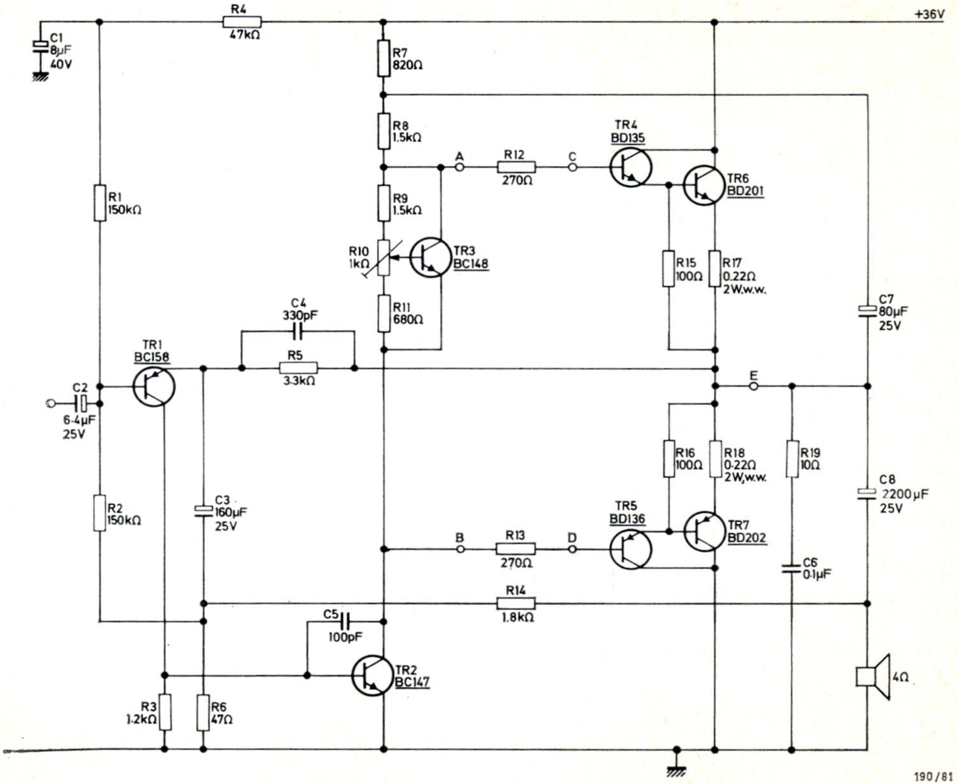

Fig. 81—25W high-quality audio amplifier circuit

25W AUDIO AMPLIFIER

The circuit of a 25W high-quality audio amplifier using BD201 and BD202 complementary transistors is shown in Fig. 81. The circuit delivers an output power of 25W into a 4Ω speaker load at better than 0·1% total harmonic distortion. The output stage uses BD201 and BD202 complementary transistors operating in class B, and BD135 and BD136 complementary transistors operating in class B are used in the driver stage. Short-circuit protection is necessary.

Circuit Description

The pre-amplifier stage uses a BC158 transistor TR_1. Feedback is applied to the emitter of this transistor by the potential-divider network formed by resistor R_6 with the parallel combination of resistors R_{14} and R_5. The pre-amplifier stage also functions as a mid-point voltage stabiliser.

The pre-driver stage uses a BC147 transistor TR_2 operating in class A. The collector load R_8 forms a bootstrap network with R_7 and capacitor C_7 to increase the gain of this stage. A capacitor C_5 is connected between the collector and base of TR_2 to limit the high-frequency gain of this transistor.

Transistor TR_3 acts as a V_{BE} multiplier to supply the base-emitter voltages of transistors TR_4 to TR_7. It also compensates for temperature variations. A BC148 transistor is used for TR_3. The quiescent current through the output transistors TR_6 and TR_7 is set by adjustment of preset resistor R_{10}.

To limit the maximum drive current through the driver and output transistors in the event of a short-circuit across the speaker terminals, resistors R_{12} and R_{13} of 270Ω are connected in series with the bases of the driver transistors. The driver stage, operating in class B, uses complementary transistors BD135 and BD136 for TR_4 and TR_5. The output stage uses BD201 and BD202 complementary transistors for TR_6 and TR_7.

The use of complementary transistors in the output stage gives a much simpler circuit than the quasi-complementary circuit usually employed in class B high-quality amplifiers. In addition, the greater symmetry of the circuit gives lower distortion, particularly in the crossover region.

Short-circuit Protection

A suitable protection circuit is shown in Fig. 82. The circuit operates on a current-trip basis and its operation is described in detail under 'Current-trip circuits' on page 153. Points of connection to the 25W amplifier are identified by the letters A, B, C, D, and E in Figs. 81 and 82.

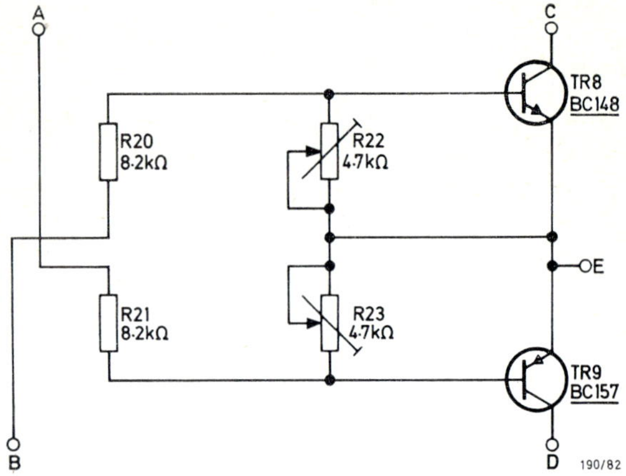

Fig. 82—Short-circuit protection circuit

Adjustment Procedure

1. Adjust the preset resistors R_{22} and R_{23} to their mid positions.
2. Connect a 2Ω load across the amplifier output terminals.
3. Connect an oscilloscope across resistor R_{17} (if a double-beam oscilloscope is used, connect the other input across R_{18}).
4. Drive the amplifier with a 1kHz signal to produce a peak voltage of about 1·5V across R_{17}.
5. Adjust R_{22} until the waveform of Fig. 83a is produced with the protection circuit commencing operation at a peak voltage of 1·4V across R_{17}.
6. The voltage across R_{18} will have reduced because the upper protection circuit is in operation. Therefore, the drive must be increased to produce a peak voltage of 1·5V across R_{18}. Then adjust R_{23} to give a waveform similar to Fig. 83b, again with protection commencing at a peak voltage of 1·4V across R_{18}.

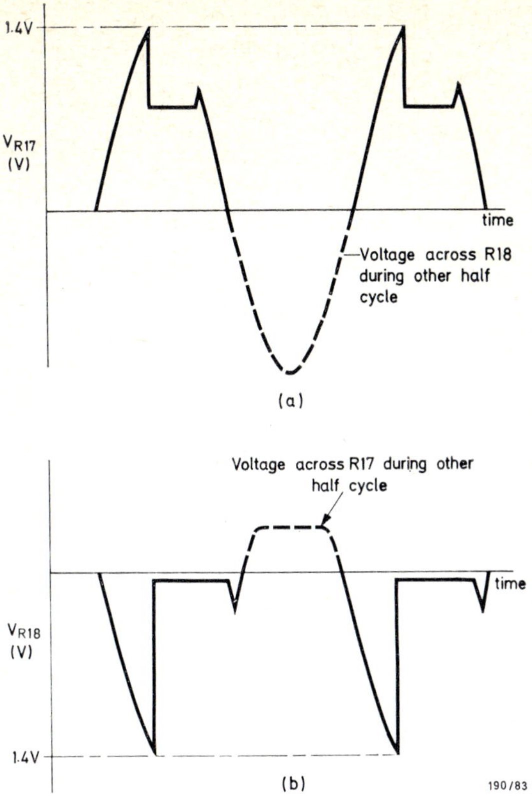

Fig. 83—Waveforms for adjustment of short-circuit protection circuit

Performance

Performance is measured at full-power output into a 4Ω load unless otherwise stated.

Full-power output	25W
Input sensitivity	400mV
Input impedance	150kΩ
Output impedance	0·05Ω
Voltage feedback factor	400 times
Unweighted signal-to-noise ratio referred to an output power of 50mW	≥76dB
Frequency response	
−1dB points at $\frac{1}{4}$ × full-power output	15Hz to 100kHz
Total harmonic distortion	see Fig. 84
Intermodulation distortion	0·5%

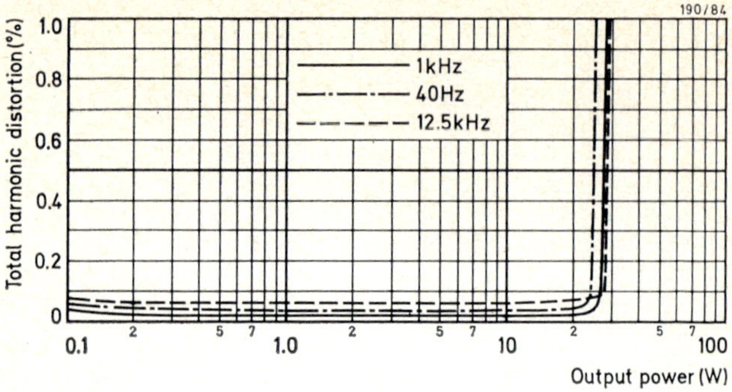

Fig. 84—Variation of total harmonic distortion with output power for three frequencies

35W AUDIO AMPLIFIER

The circuit of a high-quality 35W audio amplifier using two BD183 transistors in the output stage is shown in Fig. 85. The amplifier delivers an output power of 35W into an 8Ω load with typically 0·1% total harmonic distortion. A quasi-complementary single-ended class B push-pull configuration is used. The amplifier withstands normal overdrive conditions but short-circuit protection is necessary.

Circuit Description

The 35W amplifier circuit is similar in operation to the 25W amplifier previously described. The circuit differs only in the quasi-complementary push-pull output stage.

The driver stage uses BD139 and BD140 complementary transistors for TR_4 and TR_5. These transistors function as a phase inverter, and supply drive currents to the BD183 output transistors TR_6 and TR_7.

Short-circuit Protection

A suitable protection circuit is shown in Fig. 86. The circuit operates on a current-trip principle. Its operation is described in detail under 'Current-trip circuits' on page 153.

With an inductive load, heavy overdrive will result in the protection transistors being switched on even in the absence of short-circuit conditions. The drive is then taken from the output transistors, and if no precautions are taken the energy in the inductor will be absorbed in reverse second breakdown of a driver and an output transistor. Because this can damage the devices, two diodes D_1 and D_2 are included to prevent it occurring.

Points of connection to the 35W amplifier are identified by the letters A, B, C, D, E, F, and G in Figs. 85 and 86.

Adjustment Procedure

This circuit is adjusted in a similar manner to the 25W amplifier protection circuit.

The amplifier is driven with a 1kHz signal to produce peak voltages of about 2·2V across resistors R_{18} and R_{19}. The preset resistors are then adjusted for commencement of protection at peak voltages of 2·0V. The adjustment waveform is shown in Fig. 87.

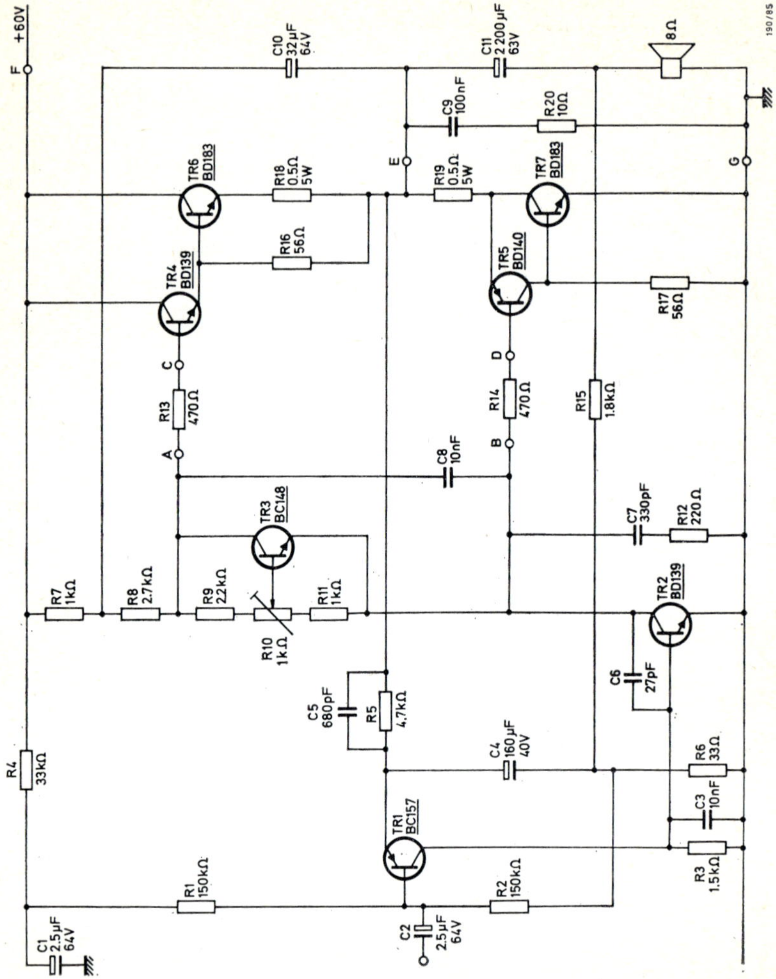

Fig. 85—35W high-quality audio amplifier circuit

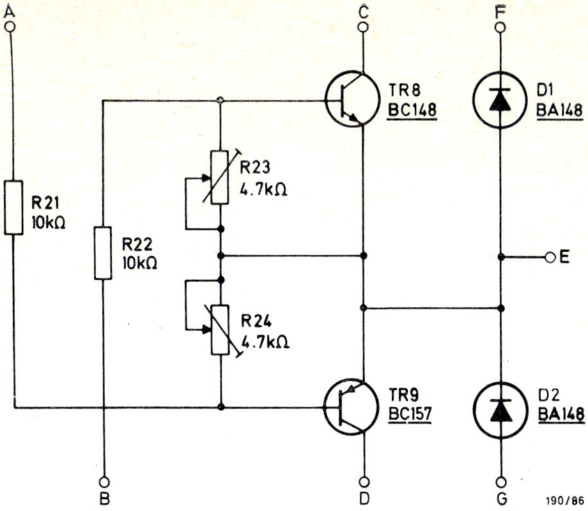

Fig. 86—Short-circuit protection circuit

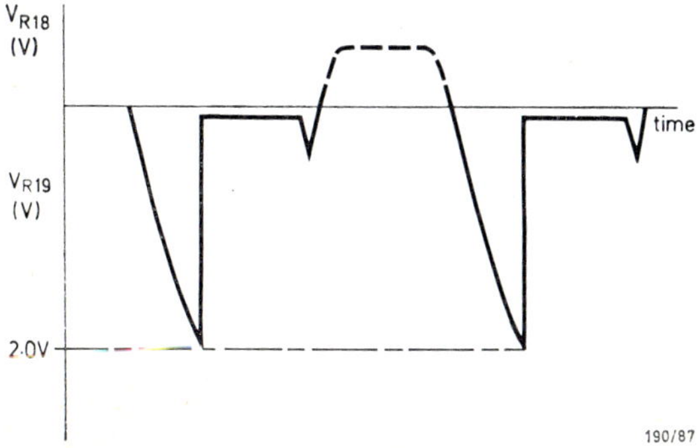

Fig. 87—Waveforms for adjustment of short-circuit protection circuit

Performance

Input sensitivity	410mV
Input impedance	150kΩ
Output impedance	0·05Ω
Voltage feedback factor	280 times
Signal-to-noise ratio (referred to an output power of 50mW, and with 1kΩ source impedance)	78dB
Frequency response −1dB points at $\frac{1}{4}\times$full-power output	10Hz to 120kHz
Intermodulation distortion	0·8%
Total harmonic distortion (with 1kΩ source impedance)	Fig. 88

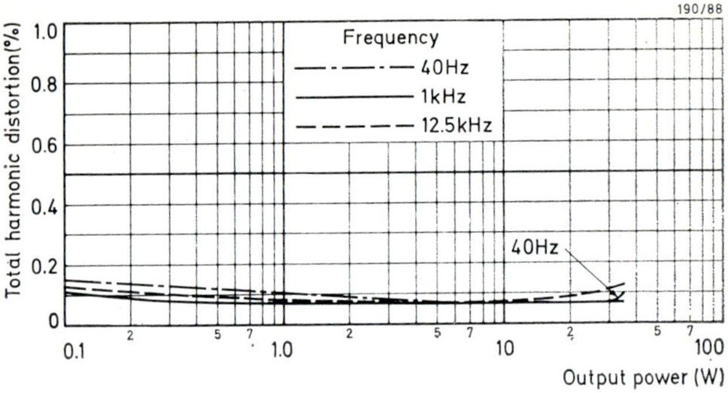

Fig. 88—Variation of total harmonic distortion with output power

50W AUDIO AMPLIFIER

The circuit of a 50W high-quality amplifier using BD184 transistors is shown in Fig. 89. The circuit delivers an output power of 50W into an 8Ω speaker load at better than 0·025% total harmonic distortion. The modified quasi-complementary output stage, operating in class B, uses a matched pair of BD184 transistors driven by complementary driver transistors BD139A and BD140A. Short circuit protection is necessary.

Circuit Description

Transistors TR_1 and TR_2 form a long-tail-pair input stage and mid-point voltage comparator, providing close control of the mid-point voltage without the need for a preset adjustment. The signal is taken from the collector of TR_1 to the pre-driver TR_3, and the output of this device is applied to the modified quasi-complementary output stage, transistors TR_5 to TR_8. Diode D_1 prevents damage to the pre-driver transistor TR_3 under short-circuit conditions. When a short-circuit occurs the output falls to zero volts, and in the absence of diode protection, transistor TR_3 would pass a current with half the rail voltage on its collector.

Transistor TR_4 compensates variation in quiescent current caused by ambient temperature and supply voltage variations. The quiescent current in the output pair TR_7 and TR_8 is set by adjustment to the preset resistor R_{12}.

The output stage differs from the conventional quasi-complementary configuration in the double bootstrapping arrangement which offers three advantages.

1. The maximum base drive to each half of the output stage is well defined, and limited to a conveniently low value. In the event of a short-circuit of the output terminals, the drive current can be shunted away from the output transistors by adding the appropriate protection circuit.
2. It allows the pre-driver to be protected as described above.
3. For a given supply voltage, a slightly higher output voltage swing is achieved.

A split power supply is used to improve the low-frequency response of the circuit.

The input capacitor C_1 is shown connected for a signal from a pre-amplifier. If a signal with respect to earth is fed to the amplifier the capacitor should be reversed so that the negative terminal is connected to the input.

Short-circuit Protection

Two suitable protection circuits are shown in Fig. 90. The first circuit (Fig. 90a) operates on a current-limit principle, and the second (Fig. 90b)

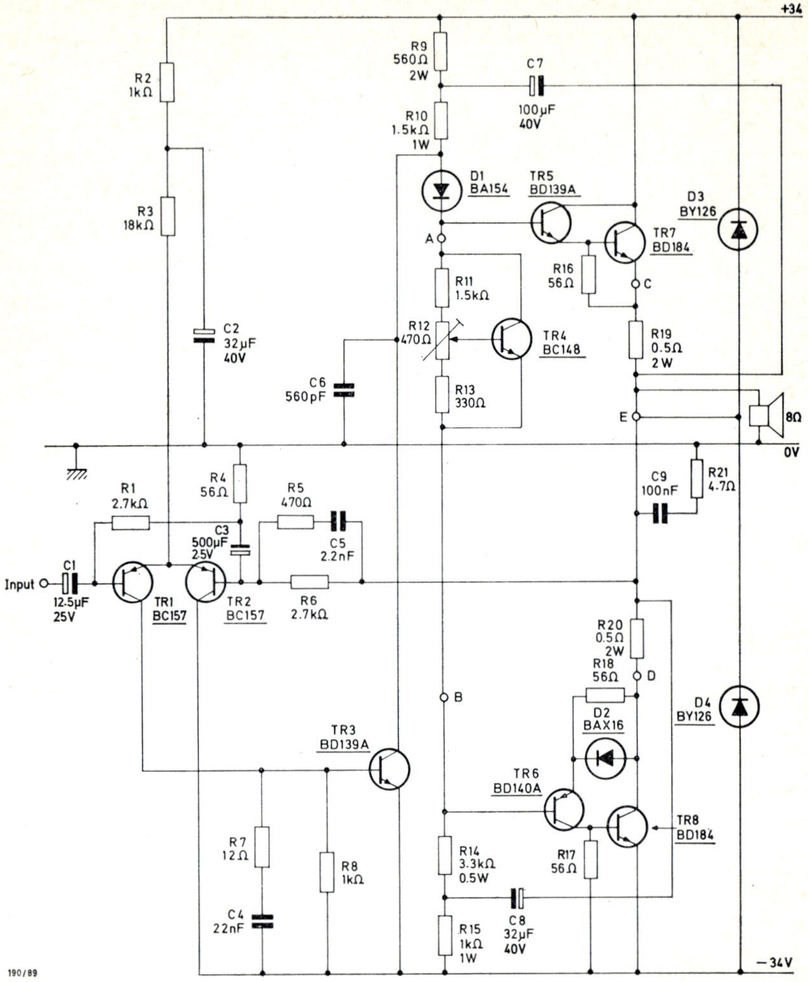

Fig. 89—50W high-quality audio amplifier circuit

on a current-trip principle. However, the first circuit gives only short-term protection. The operation of both these circuits is described with reference to the basic protection circuits on page 153. Points of connection to the 50W amplifier are identified by the letters A, B, C, D, and E in Figs. 89 and 90.

Adjustment Procedure

The current-limit and current-trip circuits should be adjusted so that they commence operating at 5A.

1. Adjust the preset resistors to their mid-positions.
2. Connect a 4Ω load across the amplifier output terminals.
3. Connect an oscilloscope across one of the emitter resistors, R_{19} or R_{20}.
4. Apply a 1kHz input signal of sufficient amplitude to produce a peak voltage of about 2·5V across the appropriate emitter resistor.
5. Adjust the appropriate preset resistor for limiting or tripping at about 2·5V.
6. Similarly adjust the other preset resistor following steps 3 to 5.

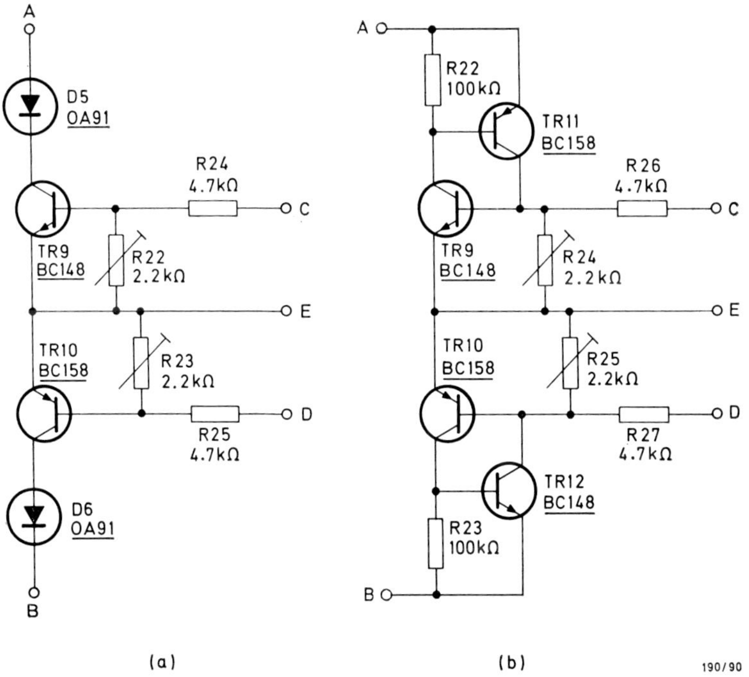

Fig. 90—Short-circuit protection circuits

Performance

Sensitivity	420mV
Input impedance	90kΩ
Total harmonic distortion at $P_0 \leqslant 50W$	
at 1kHz	<0·025%
at 12·5kHz	<0·15%
Intermodulation distortion	
just before clipping	<0·25%
for less than 5V output	<0·05%
Signal-to-noise ratio	
at 50mW output power	>76dB
Frequency response at $\frac{1}{2}\times$ full-power output,	
−1dB points	10Hz to 20kHz
−3dB points	7Hz to 35kHz
Power bandwidth	
for 1% total harmonic distortion	<20Hz to >20kHz

PRE-AMPLIFIER FOR USE WITH 10W AUDIO AMPLIFIER

A pre-amplifier for use with the 10W audio amplifier is shown in Fig. 91.

Circuit Description

Transistors TR_1 and TR_2 form a directly coupled pair, the base voltage of TR_1 being derived from the emitter of TR_2. Equalisation for tape head, magnetic pick-up, ceramic pick-up and radio inputs is obtained by feedback from the collector of TR_2 to the emitter of TR_1. The equalisation characteristics are shown in Figs. 92 and 93.

No value is given for R_2, the input resistor for the radio position, because this must be chosen to suit the available signal*. The sensitivity at the base of TR_1 is 7·5mV for 200mV output.

Transistor TR_3 is used in a feedback tone control circuit which gives a voltage gain of three times. A circuit with linear potentiometers has been used because more boost can be obtained, and also because of the difficulty in obtaining and matching reverse-logarithmic potentiometers. A BC149 or BC109 is used because a low-noise device is required in the first stage after the volume control. The tone control characteristics are shown in Fig. 94.

The working point of TR_3 is fixed by R_{23}, R_{24} and R_{26}, a.c. feedback being removed by decoupling capacitors, C_{17}, C_{18} or C_{19}. By switching these capacitors, bass cut is obtained. Networks can also be switched in by S_E to give treble cut.

Performance

Sensitivity at 1kHz for 400mV output:	
ceramic pick-up	170mV
magnetic pick-up	4mV
tape head with tape speed of $7\frac{1}{2}$in/s	6·5mV
Total distortion for 200mV output with volume control at maximum	$< 0{\cdot}02\%$
Total distortion with volume control at minimum, measured at the volume control	see Fig. 95
Signal-to-noise ratio for 200mV output, for all inputs	$>$60dB
Total current	3mA

---

*$R_2 \approx 22k\Omega \times (V_{radio} - 7{\cdot}5)/7{\cdot}5$, where V_{radio} is the nominal output voltage of the radio (mV).

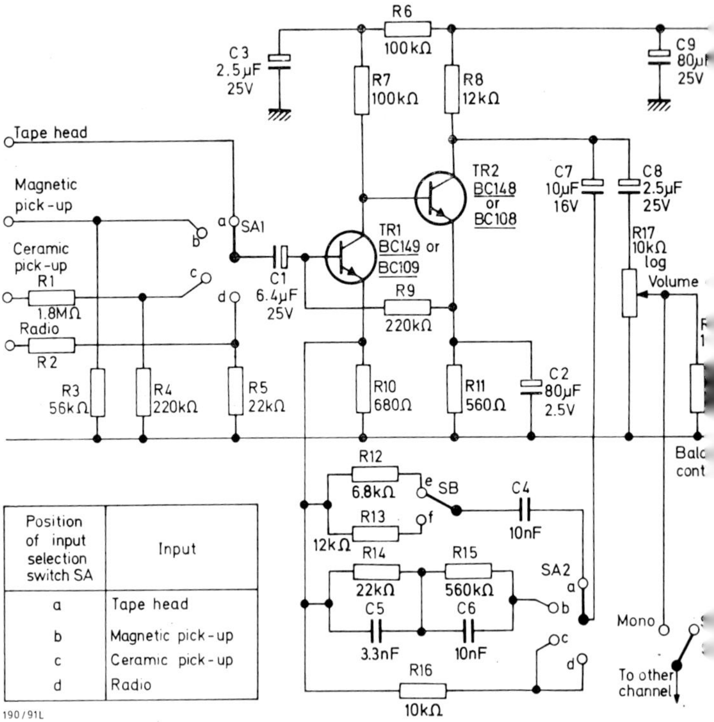

Position of input selection switch SA	Input
a	Tape head
b	Magnetic pick-up
c	Ceramic pick-up
d	Radio

Fig. 91—Pre-amplifier

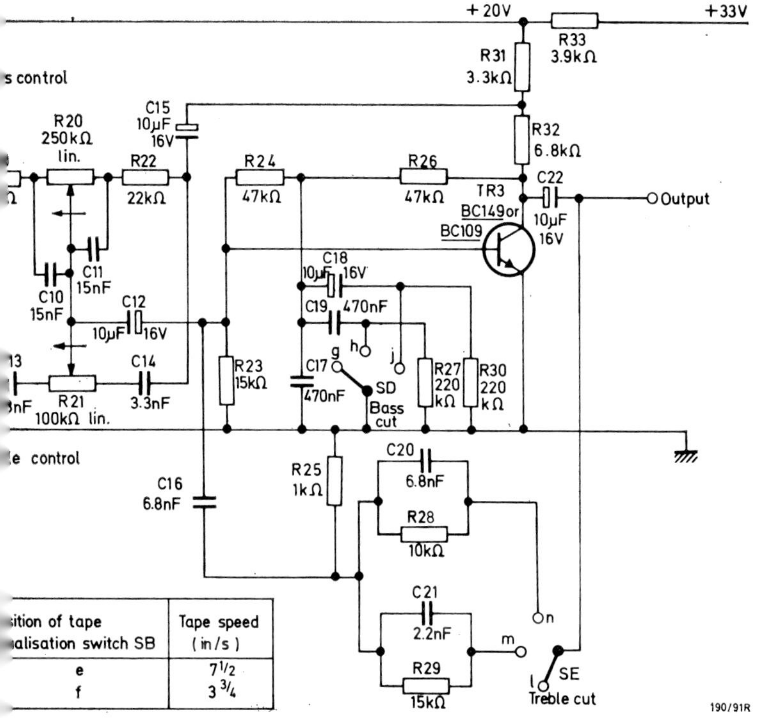

sition of tape alisation switch SB	Tape speed (in/s)
e	$7^1/2$
f	$3\,^3/_4$

with 10W amplifier

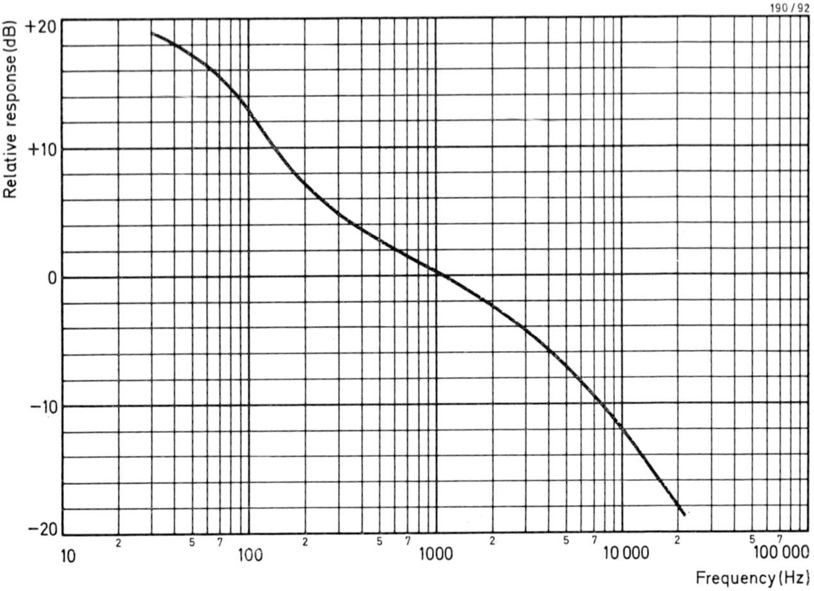

Fig. 92—Equalisation of pre-amplifier with magnetic pick-up input

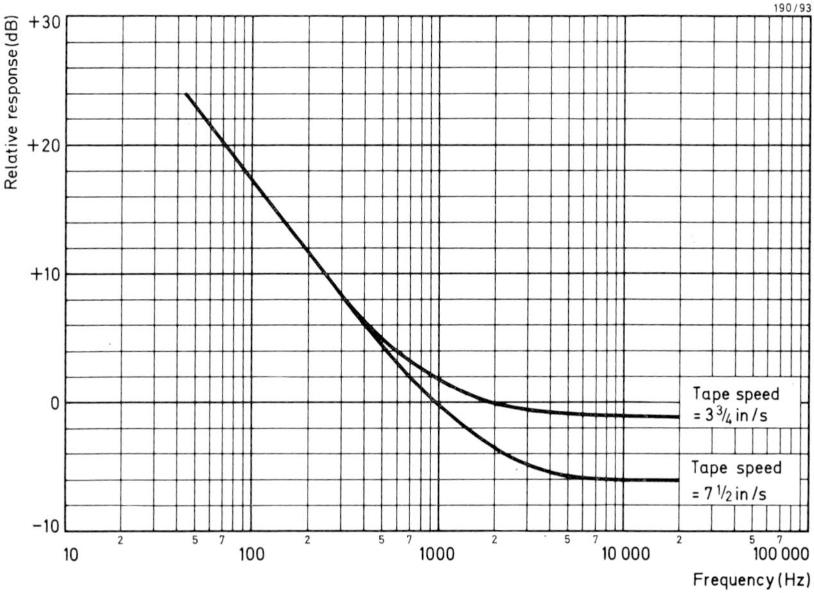

Fig. 93—Equalisation of pre-amplifier with tape head input

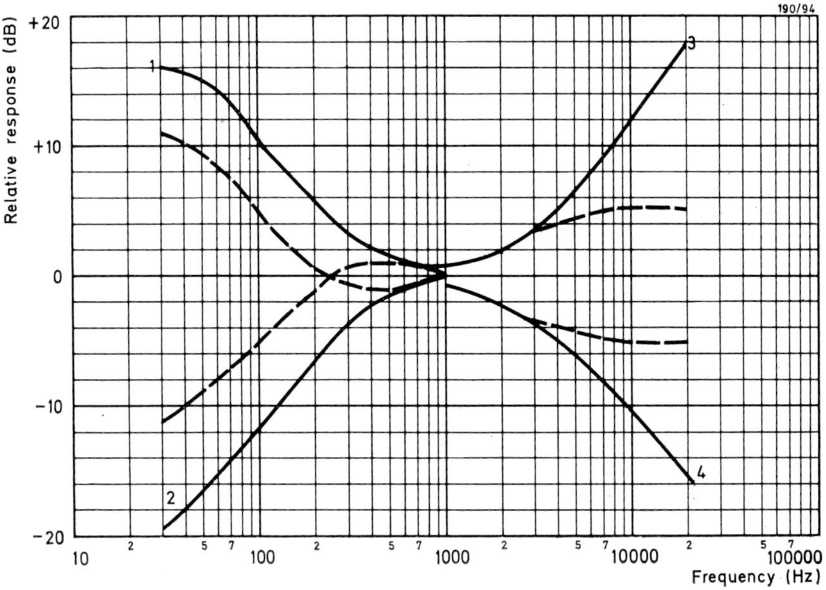

Fig. 94—Tone control characteristics of pre-amplifier
Curve 1: maximum bass boost, treble flat
Curve 2: maximum bass cut, treble flat
Curve 3: maximum treble boost, bass flat
Curve 4: maximum treble cut; bass flat
The broken curves represent intermediate positions of the controls

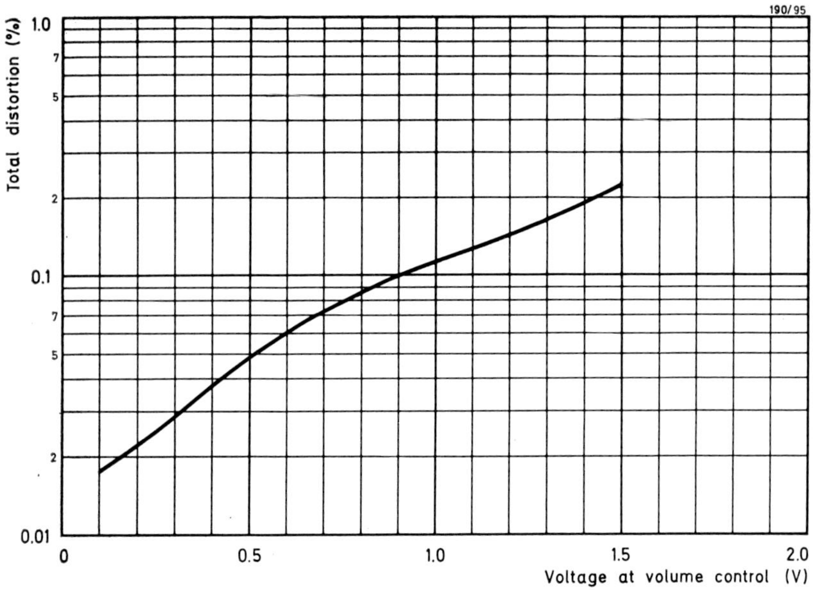

Fig. 95—Total pre-amplifier distortion (measured at the volume control) plotted against the voltage at the volume control

PRE-AMPLIFIER FOR USE WITH 15/20W, 25W, 35W, AND 50W AUDIO AMPLIFIERS

A universal pre-amplifier for use with the 15/20W, 25W, 35W, and 50W amplifiers previously described is shown in Fig. 96. The circuit uses four transistors, and may be used with power amplifiers of about 400mV sensitivity. High a.c. and d.c. feedback ensures a very low distortion level (less than 0·15%) and minimises the effects of component parameter spreads. The output impedance is less than 1kΩ and the −1dB frequency response, with tone controls in the flat position, is 10Hz to 50kHz. Bass and treble controls can be adjusted independently, to give between ±16dB and ±20dB variation.

The circuit has five inputs for magnetic and crystal pick-ups, radio tuner, tape recoder, and magnetic microphone. The sensitivity in each position is given in the performance figures. An output signal for tape recording is also available.

The pre-amplifier may be used in stereo installations. The circuit diagram shows the left-hand channel, with points of connection to the right-hand channel.

Scratch and rumble filters can be incorporated into the pre-amplifier. Suitable circuits are described on page 186.

Circuit Description

The two input transistors TR_1 and TR_2 are low-noise types BC149C and BC149B to reduce the noise output of the first two stages. The stages are directly coupled, with d.c. feedback through resistor R_{11} to stabilise the transistor d.c. conditions. A degree of a.c. feedback, varying for the different input sources, is also applied over these stages.

In the magnetic pick-up input circuit the first two stages are compensated in accordance with the R.I.A.A. standard by means of the feedback network R_{16}, C_2, C_5. The remaining inputs have a frequency-independent feedback factor and flat frequency response. The overdrive margin of the first two stages is greater than 20dB referred to 350mV across the volume control.

The voltage amplifications at 1kHz from the input to the top of the volume control R_{19} are:

magnetic pickup	87·5 times
crystal pickup	1·16 times
radio tuner	2·33 times
tape recorder	1·16 times
magnetic microphone	100 times

With nominal input signals at 1kHz, the voltage across the top of the volume control is 350mV. This signal is fed through a 1MΩ resistor R_{17} to

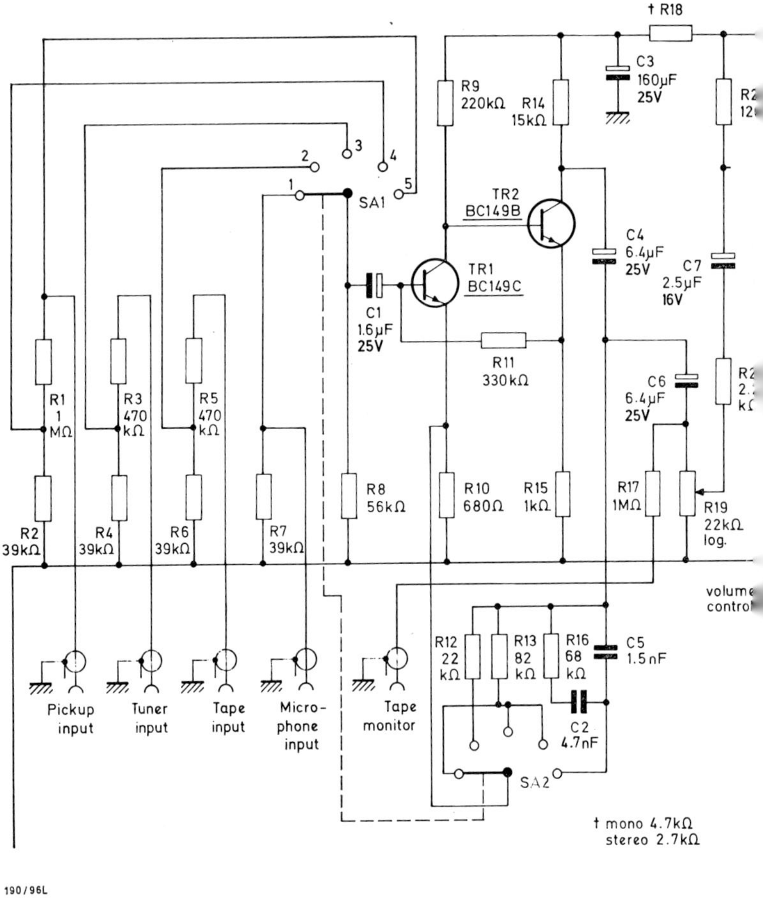

Fig. 96—Universal pre-amplifier circuit fo

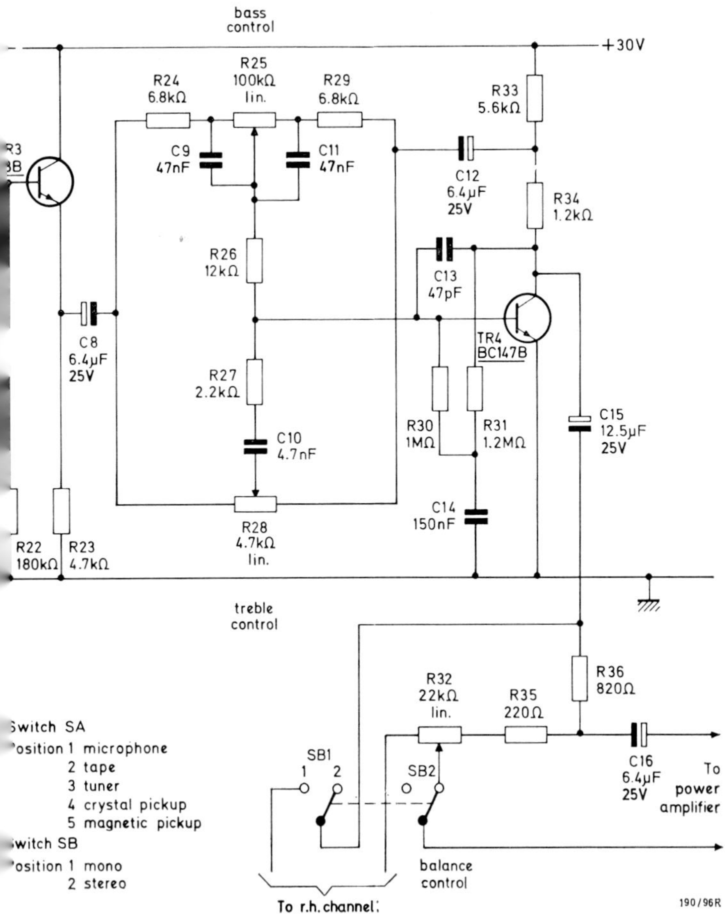

Switch SA
Position 1 microphone
2 tape
3 tuner
4 crystal pickup
5 magnetic pickup
Switch SB
Position 1 mono
2 stereo

15/20W, 25W, 35W, and 50W audio amplifiers

a tape recorder output so that a nominal recording signal of 350μV per 1kΩ of tape recorder input impedance is available.

The slider of the volume control is connected to the base of transistor TR_3, type BC148B, which operates as an emitter follower. In this configuration the transistor output impedance is low, and the tone control circuit is fed from a voltage source and so is independent of the position of the volume control slider.

The bass and treble controls are connected in the collector-to-base feedback loop of output transistor TR_4, type BC147B. This feedback system gives a low transistor output impedance and allows the tone controls to be varied independently over wide ranges. When the sliders of the tone controls R_{25} and R_{28} are in the flat position, the feedback is virtually independent of frequency and the voltage gain of the transistor is 1·35 times. The output signal from TR_4 is taken through resistors R_{36} and R_{35} and through the balance control R_{32}. The pre-amplifier output is taken from the junction of R_{35} and R_{36}.

With the slider of the balance control in the centre position, the decrease in voltage amplification is about 0·6dB. For stereo operation the slider is taken to earth. For mono operation the slider is not earthed, and the inputs of both power amplifiers are connected to the pre-amplifier output. Resistors R_{32} and R_{35}, and switch SB, can be omitted for single-channel operation.

Performance

The input sensitivity, input impedance, −1dB frequency response, and signal-to-noise ratio are given for each input. Sensitivity and input impedance are quoted for a frequency of 1kHz and for an output signal level of 440mV. The signal-to-noise ratios are given for an output signal level 20dB below 440mV.

Input	Sensitivity (mV)	Input impedance (kΩ)	Frequency response	Signal-to-noise ratio (dB)
magnetic pick-up	4	47	see Fig. 97	90
crystal pick-up	300	1000	10Hz to 35kHz	80
radio tuner	150	500	10Hz to 35kHz	80
tape recorder	300	500	10Hz to 45kHz	85
magnetic microphone	3·5	22	10Hz to 65kHz	80

Undistorted output voltage	4·4V (>20dB above nominal output voltage of 440mV)
Output impedance	<1kΩ
Total harmonic distortion 20Hz to 20kHz (all inputs)	<0·15%
Tone control characteristics	see Fig. 98

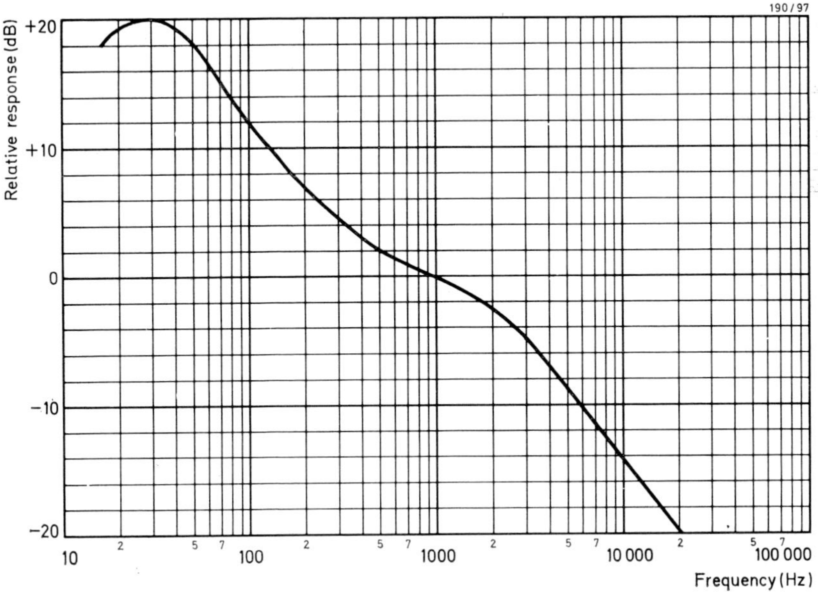

Fig. 97—Frequency response of pre-amplifier for magnetic pick-up

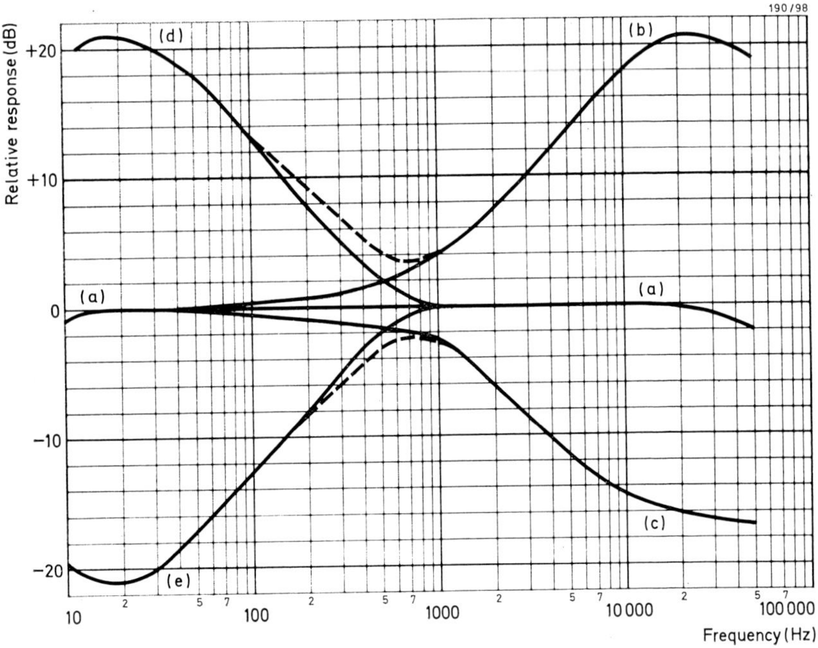

Fig. 98—Tone control characteristics of pre-amplifier
(a) bass and treble controls flat; (b) maximum treble, bass flat; (c) minimum treble, bass flat; (d) maximum bass, treble flat; (e) minimum bass, treble flat.
The broken line joining (b) and (d) indicates maximum bass, maximum treble, and that joining (c) and (e) indicates minimum bass, minimum treble.

Short-circuit Protection

In the class B amplifiers described, under operating conditions a short-circuited output will result in the output transistors drawing a current which exceeds their nominal ratings. Unless these transistors are extremely rugged they will be destroyed. Therefore, some form of protection must be included. Three methods of protection, falling in two categories, are described below.

Simple Current-limit Circuit

The basic circuit for simple current limit is shown in Fig. 99. The protection circuit is within the broken line. Transistors TR_1 and TR_2 represent a push-pull output stage. The output current through TR_1 is monitored by detecting the voltage across the emitter resistor R_1. This voltage is applied to the base of transistor TR_3 by the potential divider network R_2 and R_3. When the output current exceeds a certain value, TR_3 conducts and diverts any further base current away from TR_1. Thus, the output current through TR_1 is limited to a certain value. Diode D_1 is required to prevent reverse current from flowing through the collector-base junction of TR_3 when TR_2 is conducting. Conversely, TR_2 is protected against a short-circuit by transistor TR_4, and diode D_2 protects TR_4 when TR_1 is conducting.

The driver configuration used in this circuit prevents second breakdown of the driver transistor. If a conventional driver configuration were used instead, with a short-circuited output the driver transistor would turn hard on and cause a large current to flow through transistor TR_4. As there is already a large collector voltage on the driver under a short-circuit condition, there is a danger of second breakdown of this device.

However, in the driver configuration shown, second breakdown of the driver is prevented by the action of diode D_3. Under a short-circuit condition D_3 becomes reverse-biased and effectively disconnects the driver from the output stage, thus preventing excessive currents from damaging TR_4 and the driver transistors.

Current-trip circuits

Details are given of two basic current-trip circuits. These are shown in Figs. 100 and 101. The protection circuits are within the broken lines.

The operation of the circuit in Fig. 100 is as follows. The current through transistor TR_1 is monitored by the voltage between points E and F; that is, the sum of the voltage across R_1, the V_{BE} of TR_1, and the voltage across R_2. As the voltage between E and F increases with increasing output current through TR_1, the voltage at point G will eventually become positive with respect to point E. At a particular value this voltage will turn

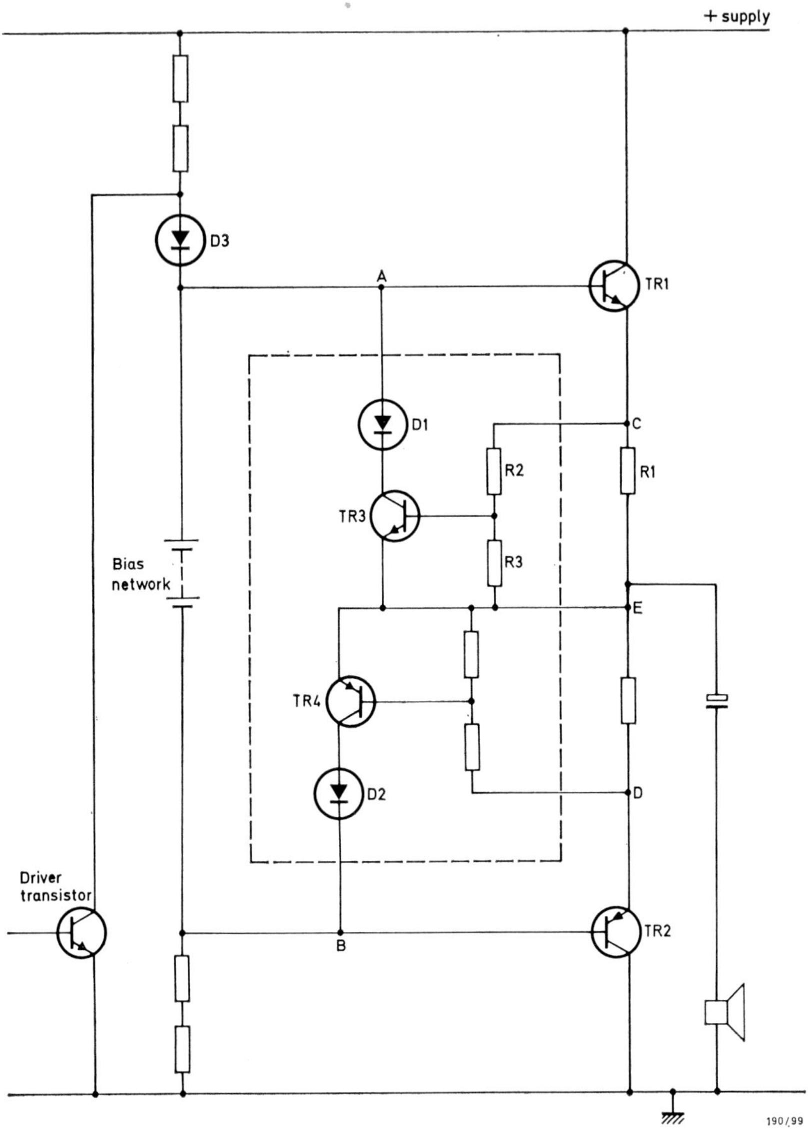

Fig. 99—Simple current-limit short-circuit protection circuit

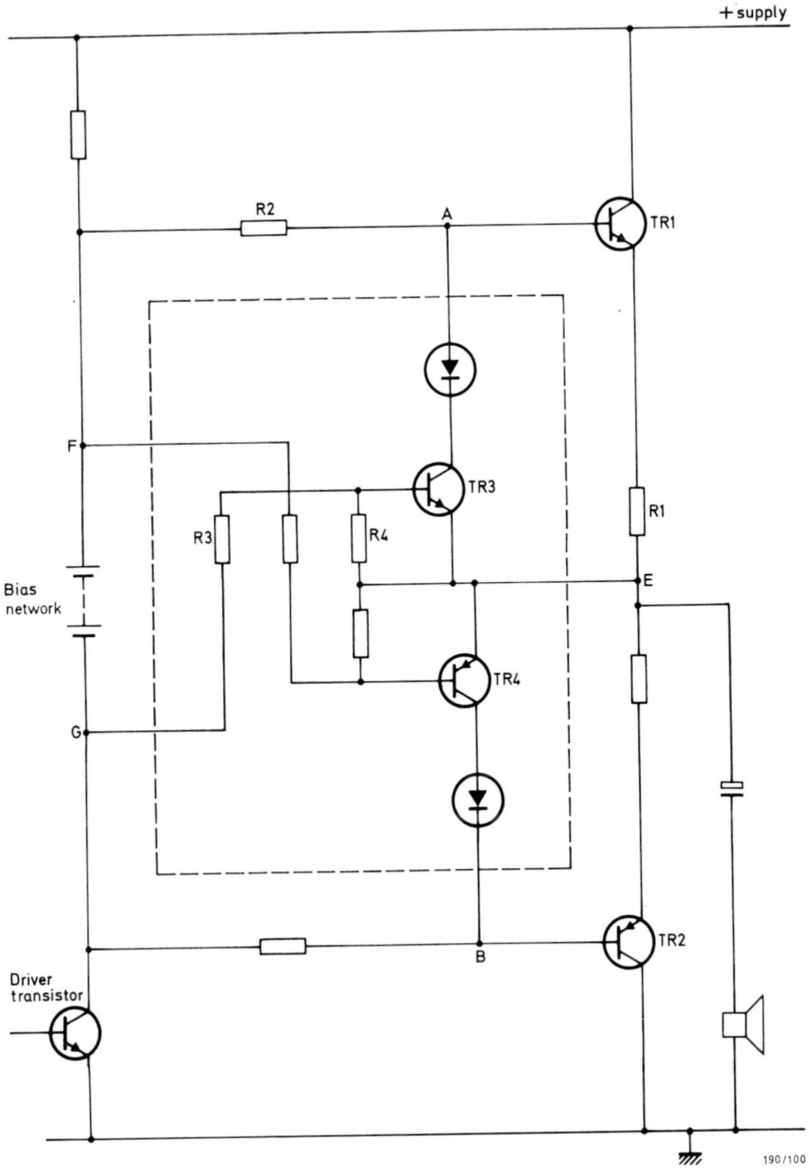

Fig. 100—Current-trip short-circuit protection circuit

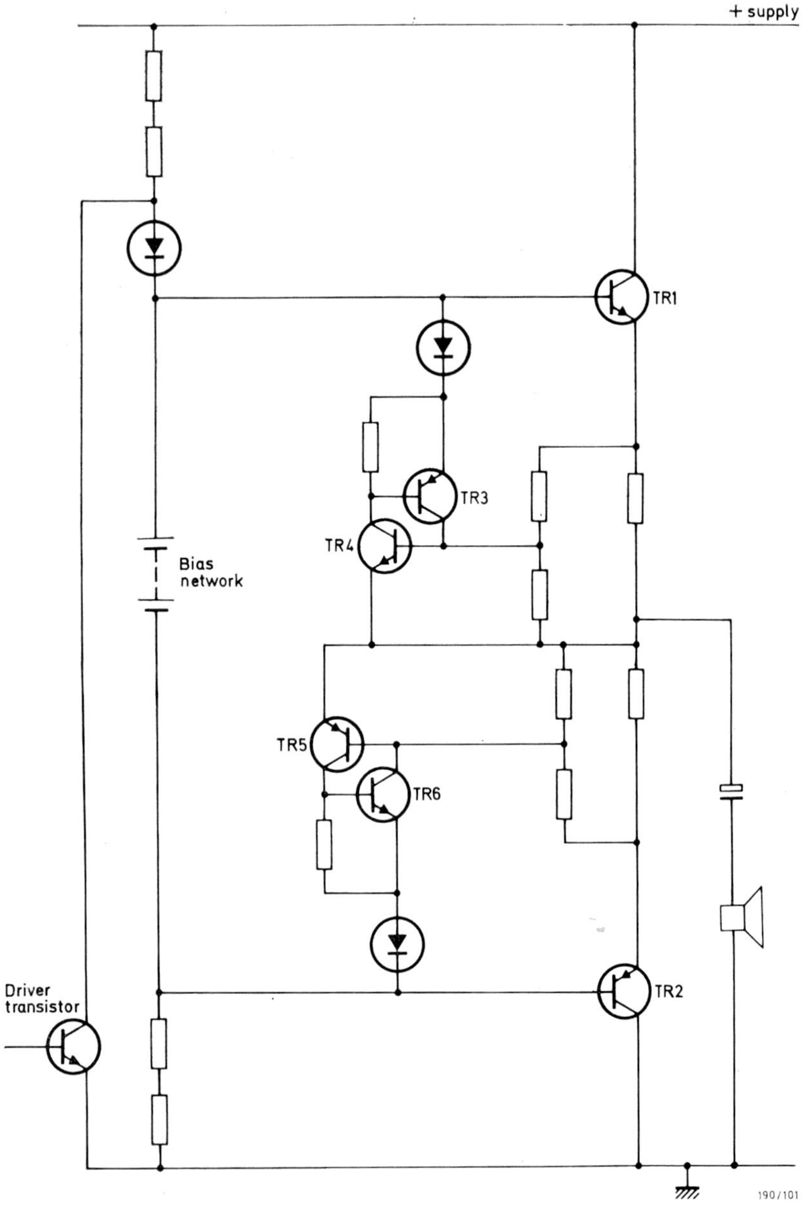

Fig. 101—Current-trip short-circuit protection circuit using 'thyristors'

on transistor TR_3 by the potential divider R_3 and R_4. The drive to TR_1 is then diverted away from the base of transistor TR_1. The harder the circuit is driven, the more TR_3 conducts, tending to turn off TR_1. Thus short-circuit protection is effected by reducing the current through TR_1 at a predetermined level. Depending on the drive and load conditions, this current could be reduced to zero. Conversely, TR_2 is protected by the action of transistor TR_4.

The circuit shown in Fig. 101 is an alternative current-trip circuit. The circuit is similar to that of Fig. 100 except that two transistor pairs, TR_3 and TR_4, TR_5 and TR_6, are connected as 'thyristors' to protect the output transistors TR_1 and TR_2.

When the current in transistor TR_1 reaches a certain limit the quasi-thyristor (TR_3, TR_4) is triggered, thus removing all base drive from TR_1. The thyristor is automatically reset during the next half-cycle. Conversely, output transistor TR_2 is protected by the quasi-thyristor formed by transistors TR_5 and TR_6.

LAYOUT OF HIGH-QUALITY AUDIO CIRCUITRY

The circuits described in this chapter are capable of high-quality reproduction, but only if they are carefully laid out and constructed. The relative positions of input, output and power supply, and the earthing and screening arrangements are of particular importance, and it is suggested that layouts should be designed with reference to the code of good practice outlined here.

General Layout

It should be remembered that the input sensitivity of an amplifier is typically 3mV at 1kHz, on the magnetic pick-up position, and the output voltage is of the order of 20V, implying a voltage gain approaching 10^4. It is therefore essential to keep the output separated and screened from the input. The magnetic field from the mains transformer may cause hum, so the transformer should be as remote as possible from the input (this is dealt with more fully below).

Earthing

Currents of several amps magnitude circulate in the power supply and output stages. It is important that no wiring carrying these currents is included in the input circuit, otherwise hum or instability, due to the small but significant resistance of the wires, will result. The paths of currents in the output stage and power supply are shown in Fig. 102.

Between points A and B there is a voltage due to TR_1; between B and C a voltage due to TR_2; and between C and D a voltage due to the power supply. The input can be earthed at A, but not at B, C, or D. In practice it is usual to combine A, B, C and D in one common earthing point.

The earthing arrangement for stereo amplifiers is considerably more complex than for mono, since the single power supply and common earth for the two signal inputs make it more difficult to avoid earth loops. The arrangement recommended is shown in Fig. 103.

Point E is the common earthing point. The voltage across the wire AE is effectively in series with the input to the power amplifier. This does not cause any trouble provided the wire is short, because the sensitivity at the input of the power amplifier is 200mV and the output-stage current is taken to the earth point separately.

Magnetic Fields

The principal source of magnetic fields in an amplifier is the mains transformer. Hum due to this cause will be at a frequency of 150Hz because the hysteresis of the core results in the magnetic field being pre-

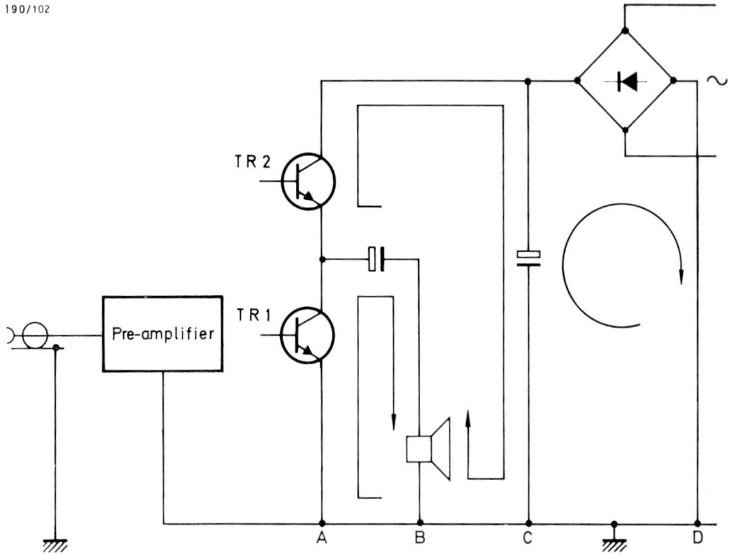

Fig. 102—Current paths in power supply and output stage

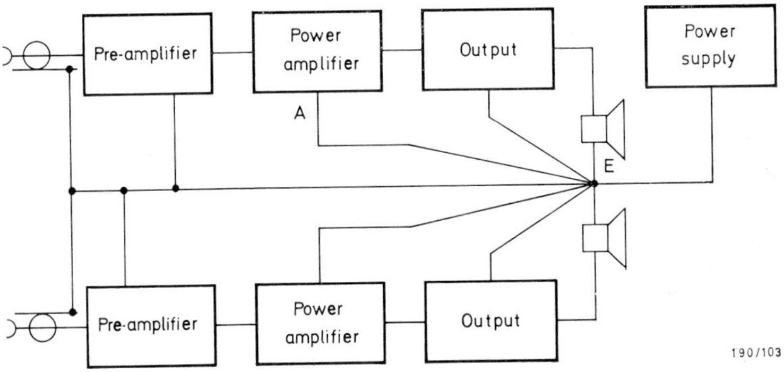

Fig. 103—Earthing arrangement for a stereo amplifier

dominantly third harmonic. The magnetic field will induce a voltage in any circuit loop.

The input loop of a pre-amplifier is shown in Fig. 104. The area of this loop should be minimised, and it may also be necessary to screen the mains transformer by making the parts of the chassis surrounding it from mild steel rather than aluminium.

The magnetic field associated with the currents in the output stage may occasionally cause trouble. Again, the remedy is to minimise the area of the circuit loops in the output stage by, for example, running emitter and collector leads close together, in order to minimise radiation.

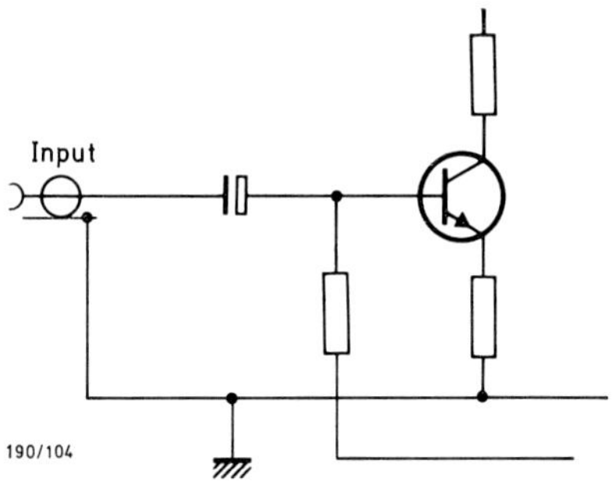

Fig. 104—Input loop of a pre-amplifier

Hum

There are many possible causes of hum and only the more probable are indicated here. A clue to the cause of the hum may often be obtained by determining its frequency.

50Hz hum may be due to stray pick-up caused by poor screening, earth loops or television break-through. The last is recognisable by the distinctive buzz of the field sync pulses, and also by the fact that it is usually accompanied by the sound channel.

100Hz hum originates in the power supply and is often indicative of poor decoupling in the h.t. line or incorrect connection of the earth leads.

150Hz hum comes from the magnetic field of the transformer and has already been discussed.

Resistor Noise

Every resistor has inherent thermal noise which could only be eliminated by refrigerating the component. However, certain resistors, notably carbon composition types, give an excess noise which is proportional to the voltage across them. It is good practice, therefore, to use high-quality resistors such as the carbon film type when building low-noise circuits. (The CR16, CR25, and CR37 series recommended for the circuits in this book are carbon film types.) Where the best low-noise performance is required, it is recommended that the collector load, base bias, and feedback resistors of input stages should be metal oxide types of not less than $\frac{1}{4}$W rating.

AUXILIARY HIGH-QUALITY CIRCUITS

In the design of the auxiliary a.f. amplifiers, control circuits and filters which follow, use has been made of the technique, common in audio engineering, of having a high input impedance and a low output impedance. All the circuits can therefore be connected together irrespective of impedance. This means that a uniform supply voltage is necessary for all the circuits, and a value of 18V has been chosen. The supply voltage can vary by $\pm 15\%$ without any appreciable change in the performance of the circuits.

AUXILIARY HIGH-QUALITY AMPLIFIERS

Basic Amplifier

The auxiliary amplifiers described in this section are derived from the basic voltage amplifier shown in Fig. 105. To achieve stabilisation against temperature variations, two d.c. feedback loops are used: one from the emitter of the second transistor to the base of the first, and the other from

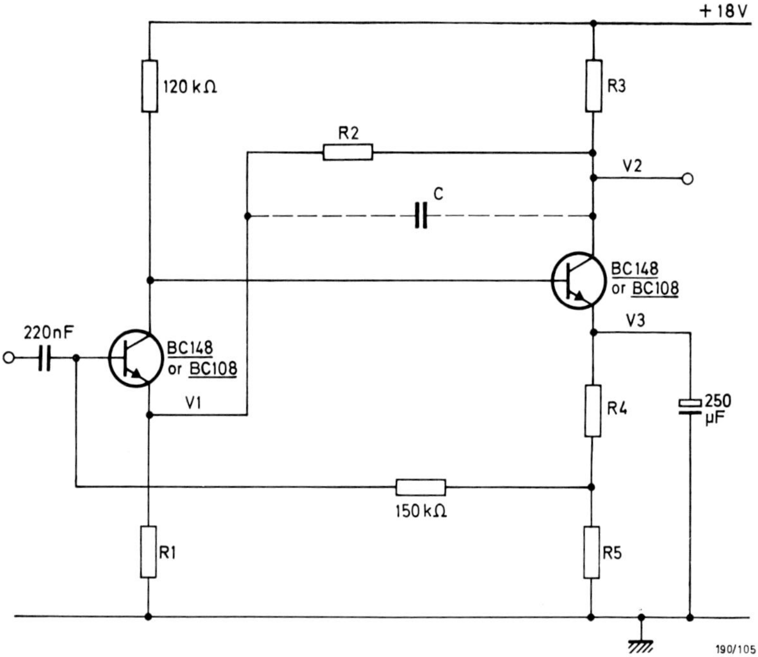

Fig. 105—Basic audio amplifier circuit. Component values are given in Table 2

the collector of the second transistor to the emitter of the first. The values of the components for circuits with voltage gains A_v of 10, 20, 30 and 40dB are shown in Table 2. The voltages in the circuit, and the input and output impedance values Z_{in} and Z_{out}, are shown in Table 3.

Figs. 106 to 109 show for the four circuits the total distortion as a function of the output voltage at three frequencies, and the noise voltage at the output as a function of the generator resistance at the input. The total distortion for all four amplifiers remains below 0·1% for output voltages up to 1V at 1kHz, and below 1% for output voltages up to 3V. The noise voltage referred to the input in all four amplifiers is less than 1μV. The frequency response (–3dB points) of all amplifiers is from 20Hz to 20kHz.

TABLE 2

Component Values

Circuit reference	A_v = 10dB	A_v = 20dB	A_v = 30dB	A_v = 40dB
R_1	4·7kΩ	1·5kΩ	1·5kΩ	1kΩ
R_2	12kΩ	15kΩ	56kΩ	180kΩ
R_3	1·8kΩ	2·2kΩ	2·2kΩ	2·2kΩ
R_4	470Ω	560Ω	330Ω	680Ω
R_5	1·2kΩ	470Ω	270Ω	220Ω
C	—	—	—	10pF

TABLE 3

Voltages and Impedances

Voltage or impedance	A_v = 10dB	A_v = 20dB	A_v = 30dB	A_v = 40dB
V_1	3·4V	0·97V	0·4V	0·15V
V_2	10·8V	9·3V	9·3V	9·7 V
V_3	5·6V	3·55V	2·3V	3·4V
Z_{in}	145kΩ	140kΩ	135kΩ	110kΩ
Z_{out}	63Ω	140Ω	260Ω	700Ω

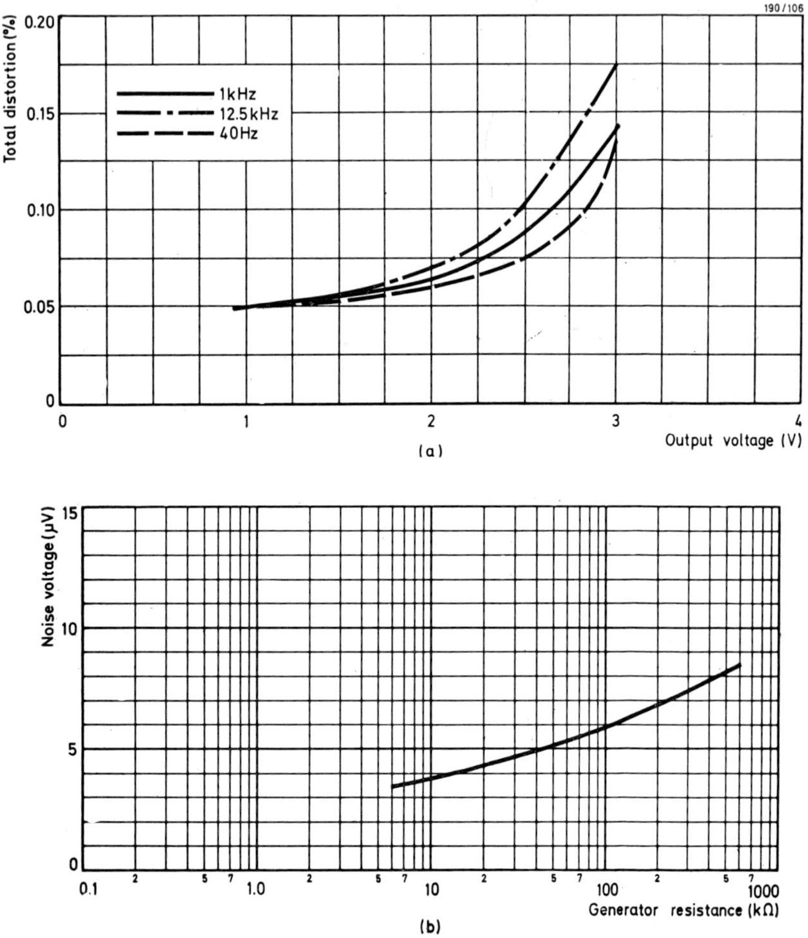

Fig. 106—Total distortion and noise voltage at output of 10dB amplifier

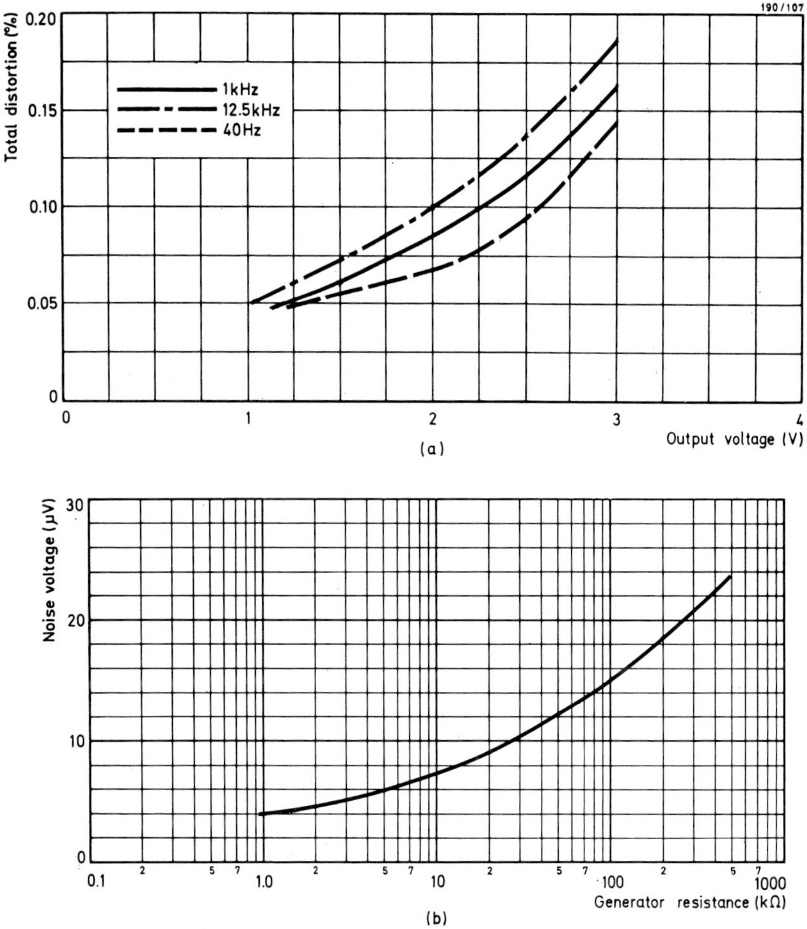

Fig. 107—Total distortion and noise voltage at output of 20dB amplifier

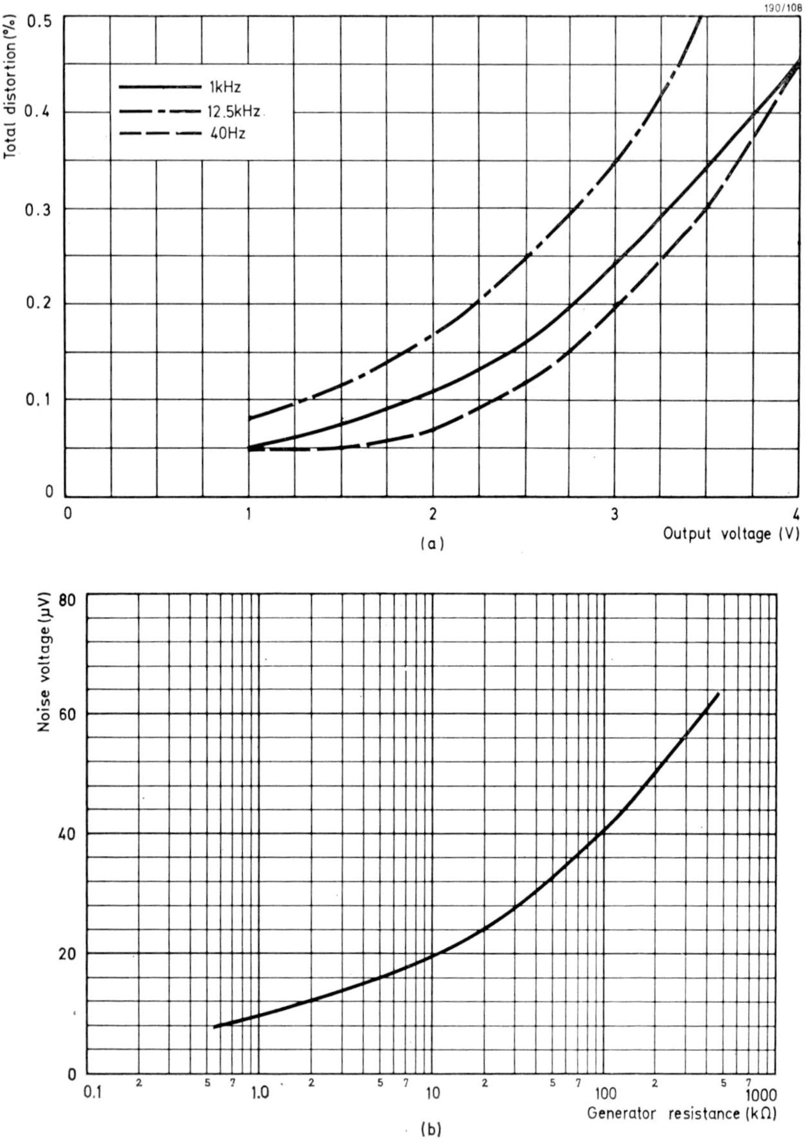

Fig. 108—Total distortion and noise voltage at output of 30dB amplifier

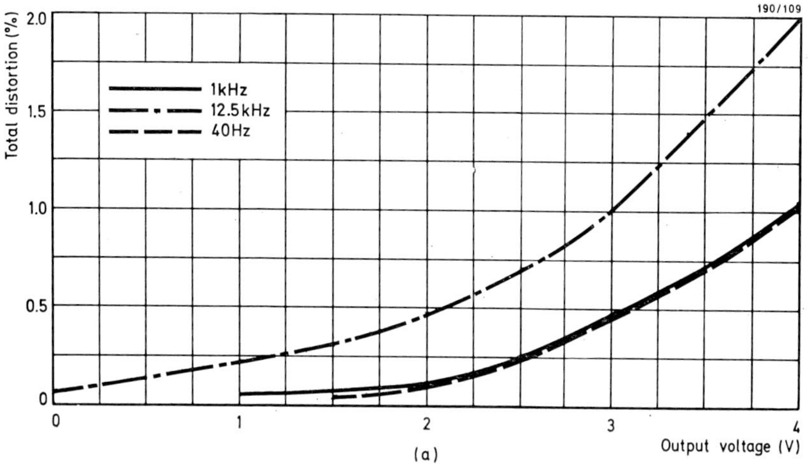

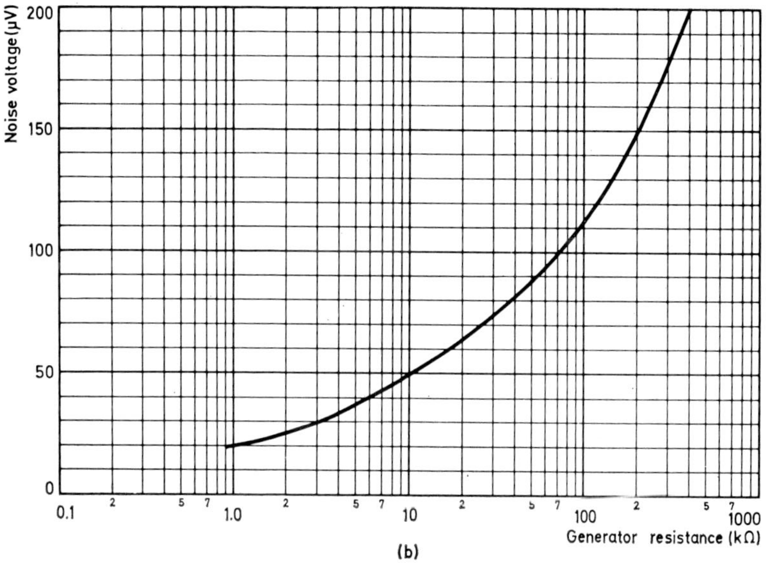

Fig. 109—Total distortion and noise voltage at output of 40dB amplifier

Buffer Amplifier

The circuit of a two-stage buffer amplifier is shown in Fig. 110. The first stage works in the common-emitter configuration with a large amount of feedback, while the second stage is an emitter follower. This ensures a high input impedance of 3·6MΩ, and a low output impedance of 250Ω. The voltage gain is unity, and the frequency response (–3dB points) is from 20Hz to 20kHz. Fig. 111 shows the total distortion and the noise voltage at the output. The distortion remains below 0·5% with output voltages up to 2·5V.

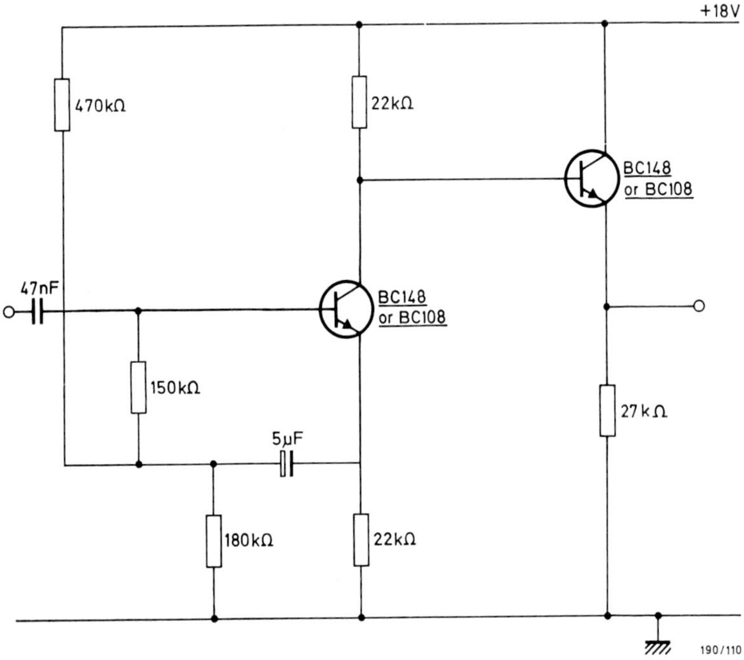

Fig. 110—Buffer amplifier circuit

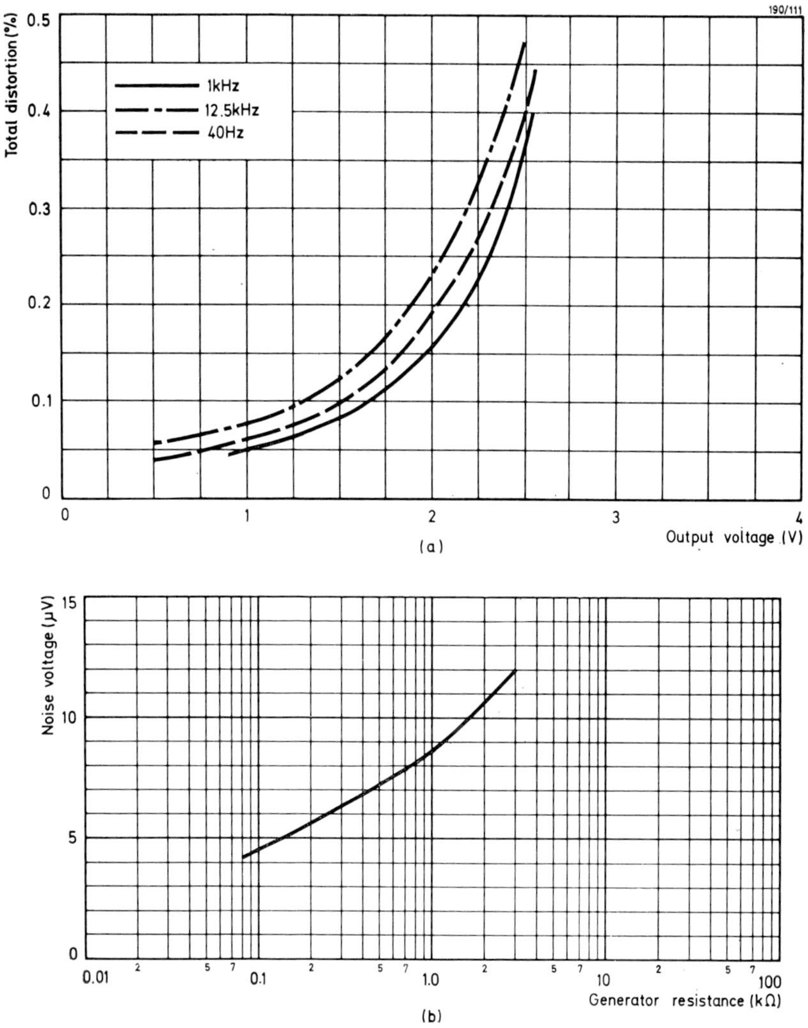

Fig. 111—Total distortion and noise voltage at output of buffer amplifier

Microphone Amplifier

The circuit of a microphone amplifier with a voltage gain adjustable between 13 and 40dB by varying the feedback is shown in Fig. 112. The total distortion for the limiting values of the voltage gain is shown in Fig. 113. With an output voltage of 2V, the distortion is 0·75% for a gain of 40dB, and 0·15% for a gain of 13dB. The values of noise voltage correspond to those of the 10dB and 40dB amplifiers of Fig. 105. The input and output impedances, and the frequency response (–3dB points) are shown in Table 4.

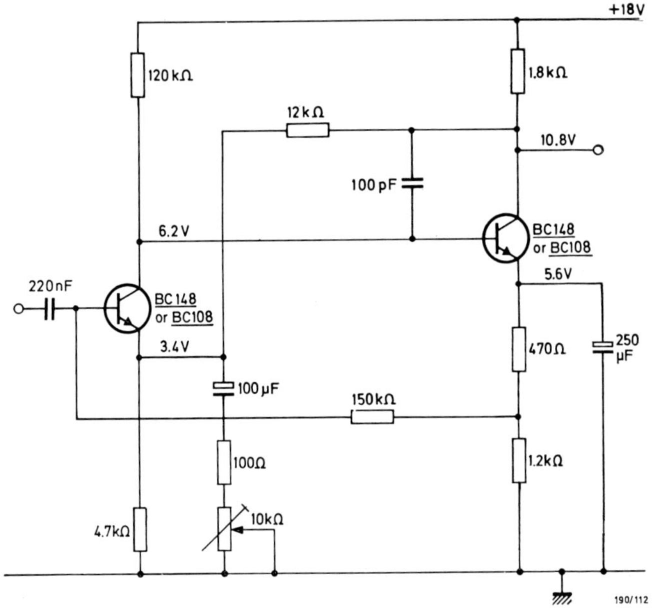

Fig. 112—Microphone amplifier circuit

TABLE 4

Impedances and Frequency Response

Impedance or frequency	A_v = 13dB	A_v = 40dB
Z_{in}	145kΩ	120kΩ
Z_{out}	47Ω	120Ω
f_{lower}	$<$ 20Hz	$<$ 20Hz
f_{upper}	$\gg$ 20kHz	$\gg$ 20kHz

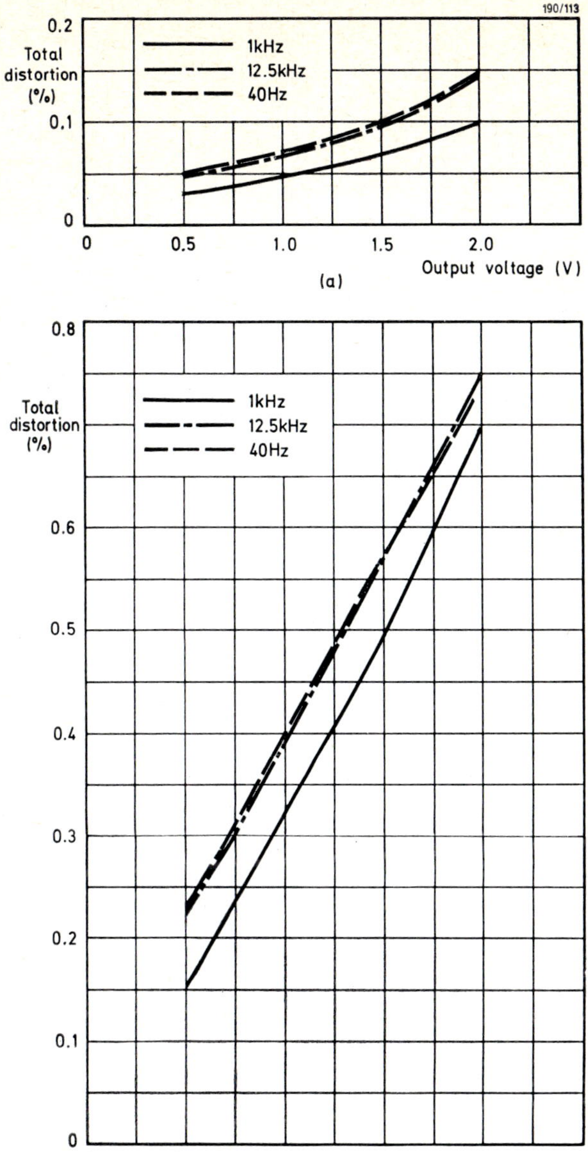

Fig. 113—Total distortion of microphone amplifier with voltage gain of (a) 13dB and (b) 40dB

Mixer Amplifier

The circuit of a 'virtual earth' mixer amplifier designed for up to (say) five inputs is shown in Fig. 114. The name 'virtual earth' derives from the action of the feedback which ensures that only very small a.c. voltages exist at the base of transistor TR_1, (that is, it is virtually an earth point). The very low impedance at the summing point gives an inherently very low inter-input crosstalk performance. The voltage gain of the mixer is R_f/R_i, where R_f is the feedback resistor and R_i the input impedance.

With a nominal output level of 350mV r.m.s., the frequency response is 30Hz to 20kHz at the −0·5dB points, and the noise is −85dB. The total harmonic distortion at nominal output and 1kHz is 0·02%, rising to 0·08% at 3V output.

The noise performance will be somewhat degraded if more than five inputs are mixed.

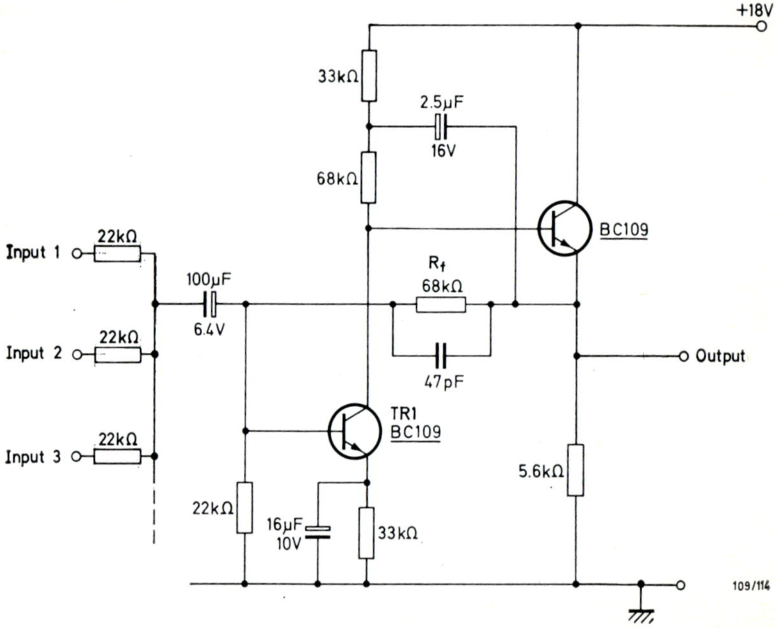

Fig. 114—Virtual-earth mixer amplifier circuit

The mixer amplifier is based on a circuit described in 'Stereo Mixer' by H. P. Walker, **Wireless World,** Vol. 77, No. 1427, May 1971.

Amplifier with High Voltage Output

The circuit of an amplifier with a voltage gain of 20dB and designed for a maximum output voltage of 10V is shown in Fig. 115. To achieve this high output voltage, it is necessary for the supply voltage to be 45V instead of the standard 18V. The total distortion at 1kHz and maximum output voltage is 0·11% (Fig. 116). The frequency response (−3dB points) is from 20Hz to 20kHz, and the input and output impedances are 140kΩ and 200Ω respectively.

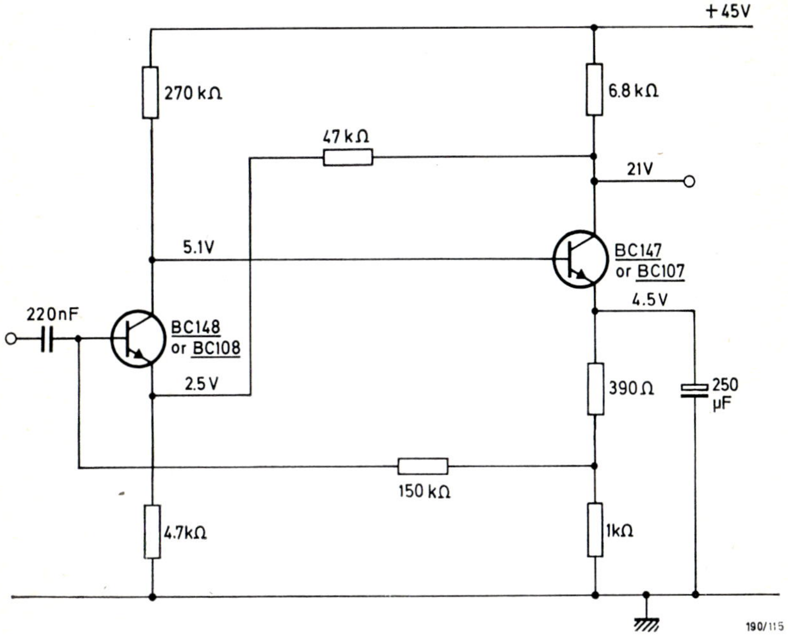

Fig. 115—Amplifier with high voltage output

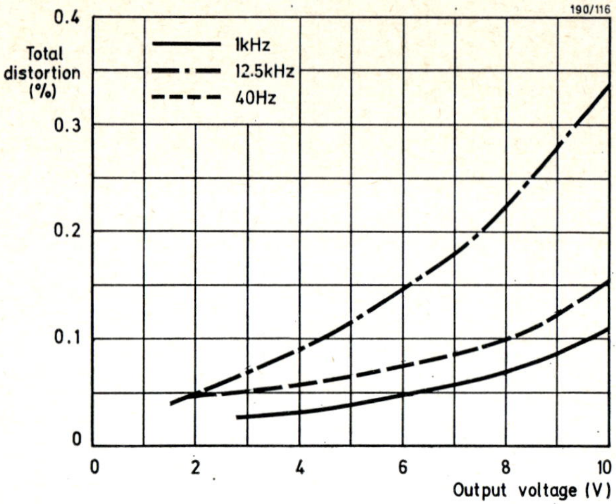

Fig. 116—Total distortion of amplifier shown in Fig. 115

AUXILIARY HIGH-QUALITY CONTROL CIRCUITS

Balance Control

The balance control of Fig. 117 makes it possible in stereophonic systems to vary the voltage gain in both channels by 6dB in opposite directions. The controlling variable resistor is inserted in the feedback circuit. The average gain is 23·4dB, and Fig. 118 shows the total distortion for minimum and maximum gain. The differences are slight because of the large amount of feedback applied. The noise voltage corresponds to that of the 20dB amplifier of Fig. 105. The frequency response (—3dB points) is from 20Hz to 20kHz, while the input and output impedances are 140kΩ and 85Ω respectively.

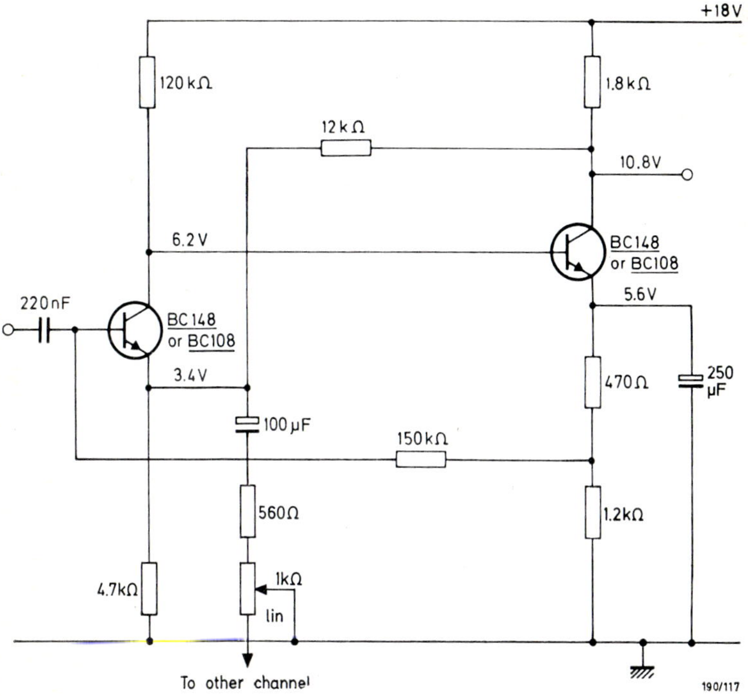

Fig. 117—Balance control circuit (only one channel shown)

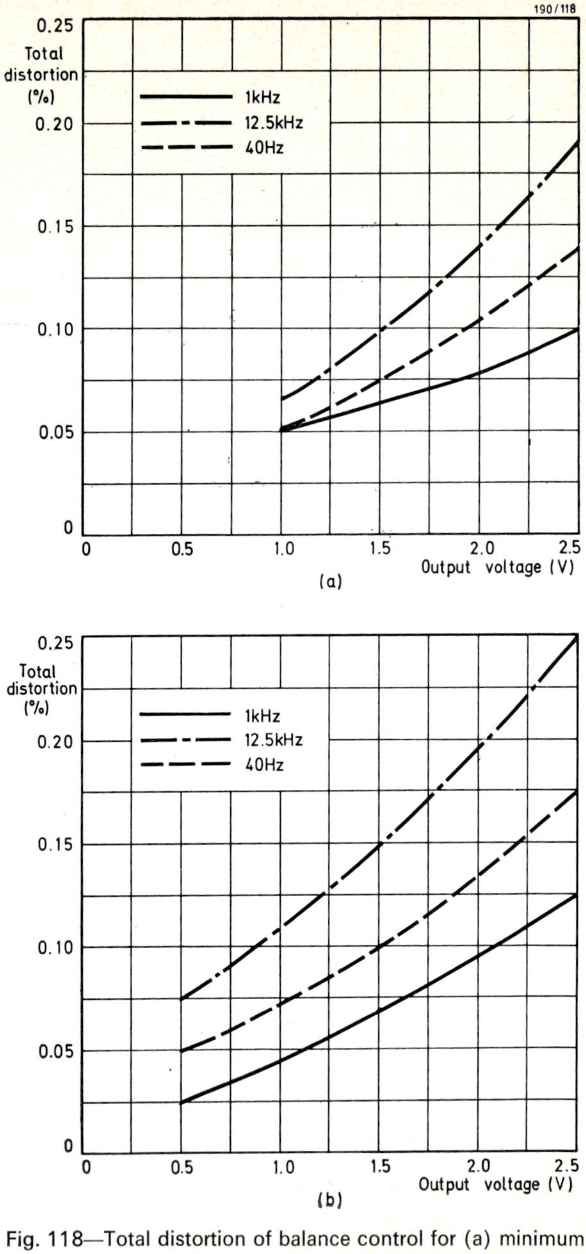

Fig. 118—Total distortion of balance control for (a) minimum and (b) maximum gain

Active Tone Control

The active tone control circuit of Fig. 119, unlike the well-known frequency-dependent voltage-divider circuits, operates with a frequency-dependent feedback network between the collector and base of the transistor. The tone control characteristics are shown in Fig. 120. The range of control extends from −22 to +19·5dB at 30Hz, and from −19 to +19·5dB at 20kHz. The flat frequency response is obtained when the variable resistors are in the physically central position. The voltage gain is then 0·91. Fig. 121 shows the total distortion as a function of output voltage for frequencies of 40Hz, 1kHz, and 12·5kHz, measured with the controls in the flat position. For small input voltages (less than 250mV), the total distortion remains below 0·1%, and for an output voltage of 2V it rises to 0·85% at 12·5kHz. The input and output impedances at 1kHz are 40kΩ and 180Ω respectively.

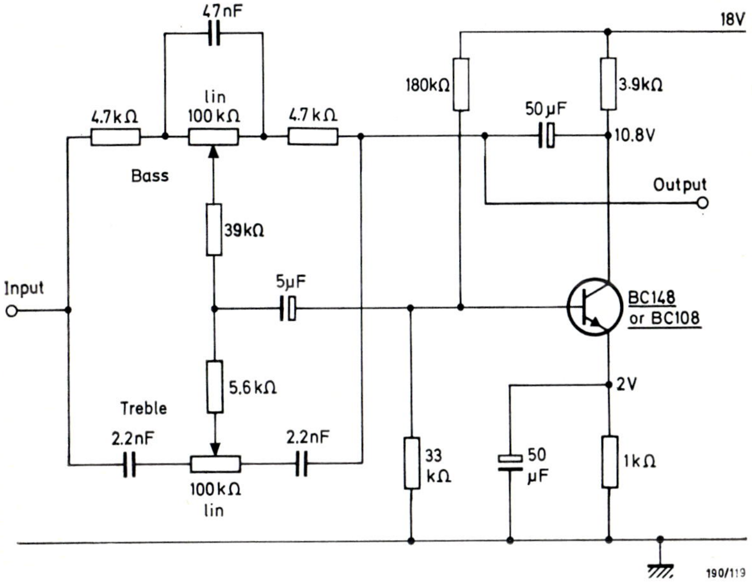

Fig. 119—Active tone control circuit

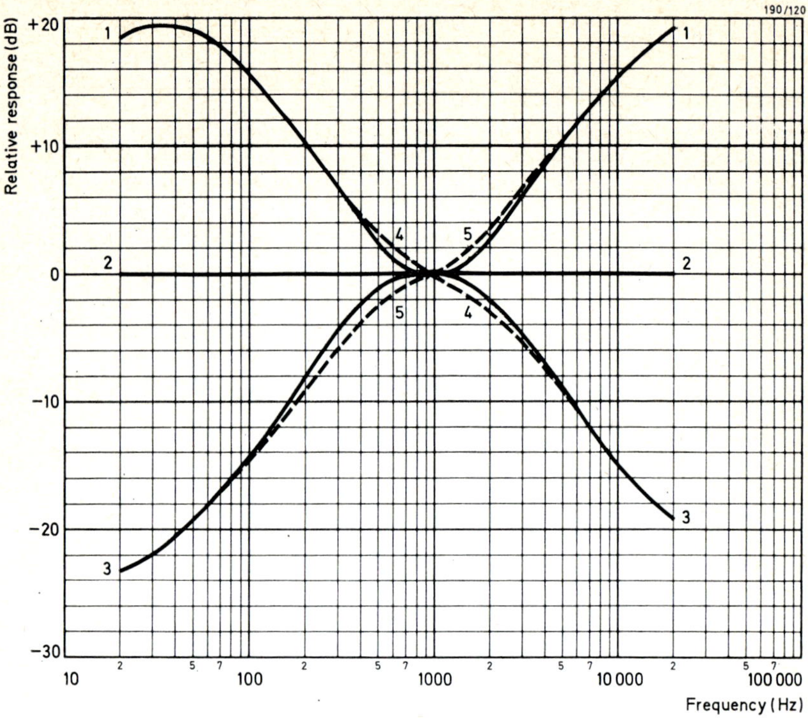

Fig. 120—Characteristics of active tone control circuit
Curve 1 : maximum bass boost, maximum treble boost
Curve 2 : linear frequency response (controls flat)
Curve 3 : maximum bass cut, maximum treble cut
Curve 4 : maximum bass boost, maximum treble cut
Curve 5 : maximum bass cut, maximum treble boost

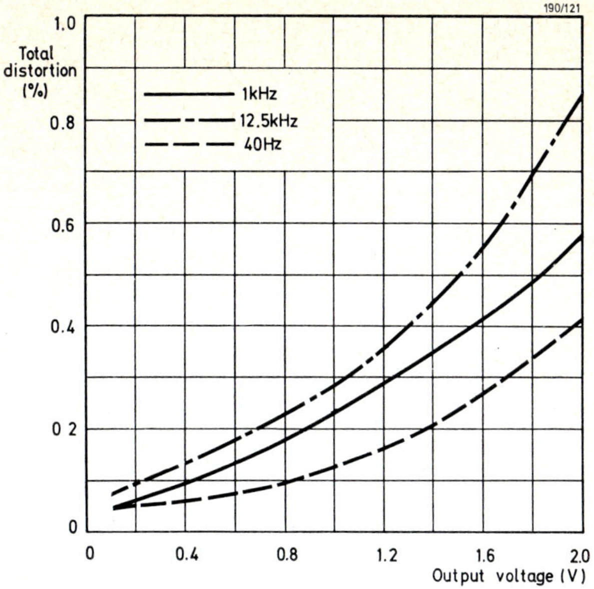

Fig. 121—Total distortion of active tone control

Sound Source Width Control

Fig. 123 shows a circuit by which the width of the sound source in a stereophonic system can be varied continuously. For this, part of the signal voltage of one channel is added to the second channel. Two ganged 10kΩ potentiometers are adjusted to control the sound source width continuously between in-phase crosstalk of 100% (corresponding to mono operation) and anti-phase crosstalk of 24%. Greater anti-phase crosstalk is not necessary since the sound impression will 'fall apart' for greater values. A 5kΩ preset potentiometer is used to balance the two channels. The voltage gain of the circuit is 0·5. Input and output impedances are 750kΩ and 47Ω respectively, and the frequency response (–3dB points) is from 20Hz to 20kHz.

Fig. 122 shows the total distortion of the circuit as a function of output voltage for three frequencies.

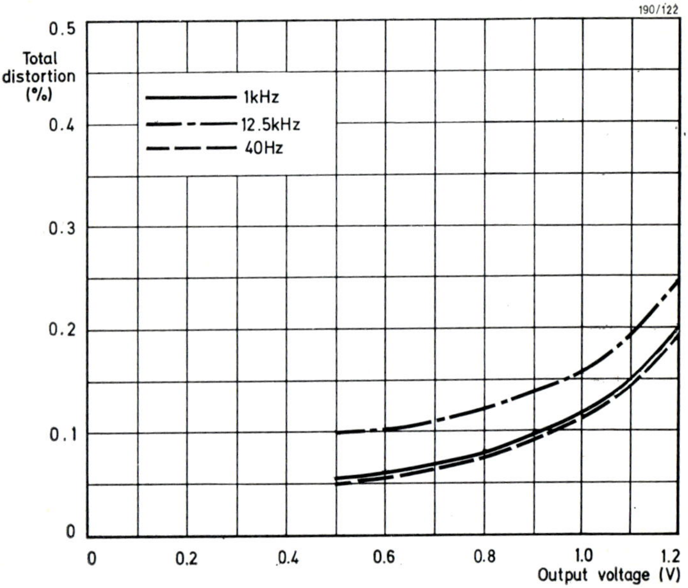

Fig. 122—Total distortion of sound source width control

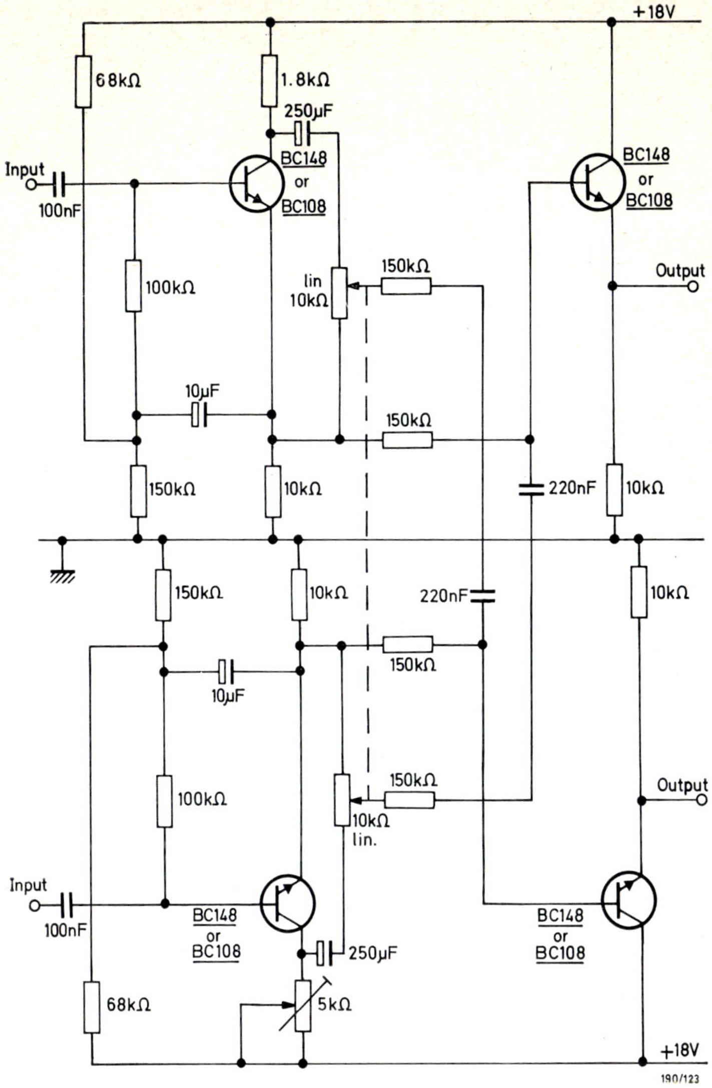

Fig. 123—Sound source width control circuit

AUXILIARY HIGH-QUALITY FILTERS

Low-pass/High-pass Filter

The circuit of a low-pass/high-pass filter is shown in Fig. 124. It consists of two RC networks connected in series with a buffer amplifier, the circuit of which is given in Fig. 110. The adjustable frequency characteristics are shown in Fig. 125. The following frequency limits can be selected:

lower frequency 40, 80, 160, and 270Hz;
upper frequency 11, 9, 4·5, and 3·2kHz.

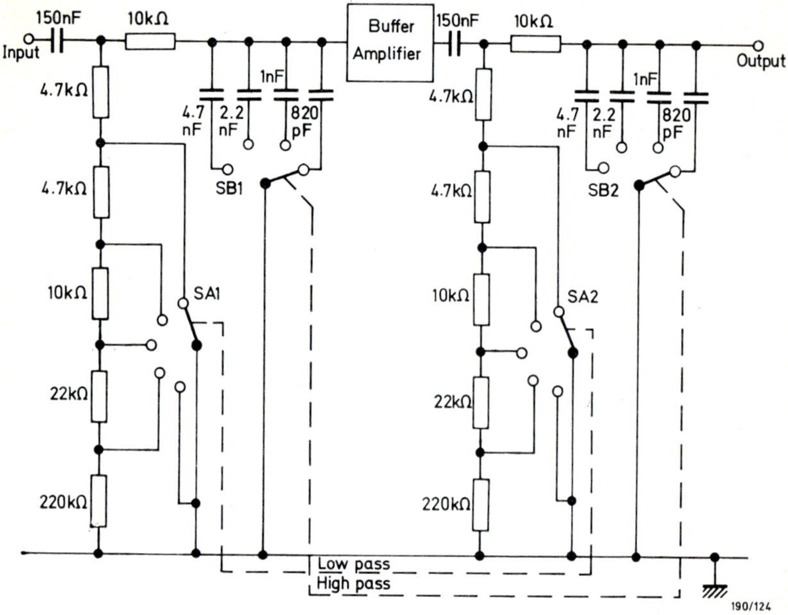

Fig. 124—Low-pass/high-pass filter circuit

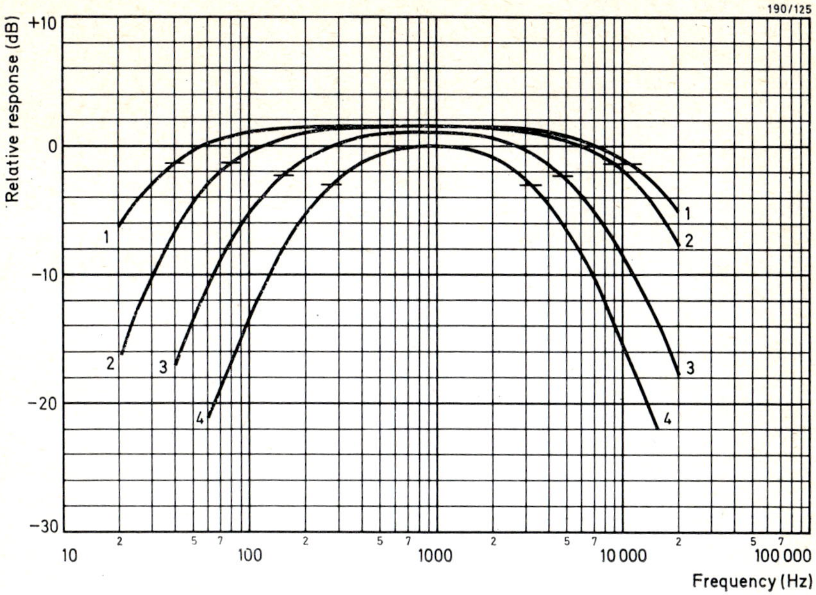

Fig. 125—Frequency characteristics of low-pass/high-pass filter
Curve 1: $f_{c1} =$ 40Hz, $f_{c2} =$ 11kHz
Curve 2: $f_{c1} =$ 80Hz, $f_{c2} =$ 9kHz
Curve 3: $f_{c1} =$ 160Hz, $f_{c2} =$ 4·5kHz
Curve 4: $f_{c1} =$ 270Hz, $f_{c2} =$ 3·2kHz
(f_{c1} = lower cut-off frequency; f_{c2} = upper cut-off frequency)

Noise and Rumble Filter

The circuit of a noise and rumble filter is shown in Fig. 126. Bass and treble cut are produced by an RC network connected between two emitter followers, and a feedback loop from the output to the input through a second RC network. In this way, a high slope of about 13dB per octave is achieved. The frequency limit of the rumble filter is fixed at 45Hz, and the noise filter can be switched to limits of 16, 12, and 7kHz. The resulting frequency characteristics are shown in Fig. 127. The voltage gain is 0·95, and the total distortion at 1kHz and an output voltage of 2V is 0·35%, falling to less than 0·1% at 1V. The input and output impedances are 1·7MΩ and 450Ω respectively.

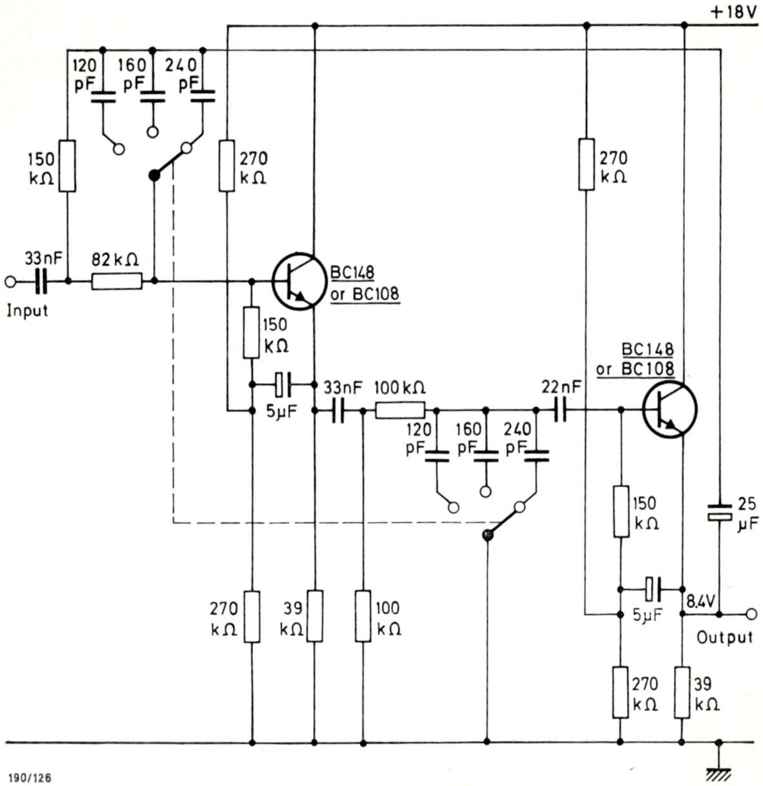

Fig. 126—Noise and rumble filter circuit

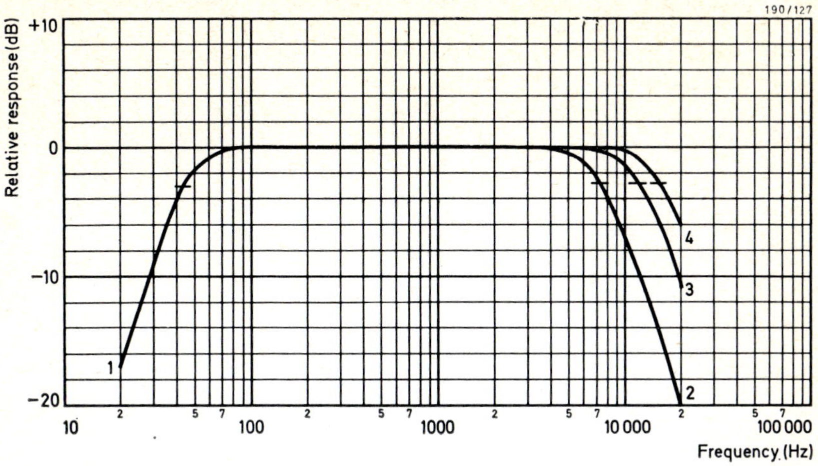

Fig. 127—Frequency characteristics of noise and rumble filter

For all curves $f_{c1} = 45Hz$
Curve 2: $f_{c2} = 7kHz$
Curve 3: $f_{c2} = 12kHz$
Curve 4: $f_{c2} = 16kHz$

Scratch and Rumble Filters for use with Universal Pre-amplifier

Scratch and rumble filters can be incorporated into the universal pre-amplifier described on page 147. The scratch filter has a roll-off of 10dB/octave, and switchable −3dB points of 17, 11, 7·5, and 5kHz. The rumble filter has a slope of 12dB/octave with a −3dB point at 40Hz. The circuits of the two filters show them incorporated in the universal pre-amplifier, and the component references apply to the pre-amplifier circuit on pages 148 and 149.

Rumble Filter

The circuit of the rumble filter is shown in Fig. 128. It is inserted between transistors TR_2 and TR_3, switch S_C reducing capacitor C_6 (6·4µF) to 220nF and C_7 (2·5µF) to 100nF. As the volume control is part of this network, the operation is affected slightly by the setting of the control. Between half-volume (6dB down) and full volume, the −3dB point rose from 38 to 42Hz. At 20dB down, the −3dB point was 37Hz. The filter characteristics for outputs of 44 and 440mV are shown in Fig. 129.

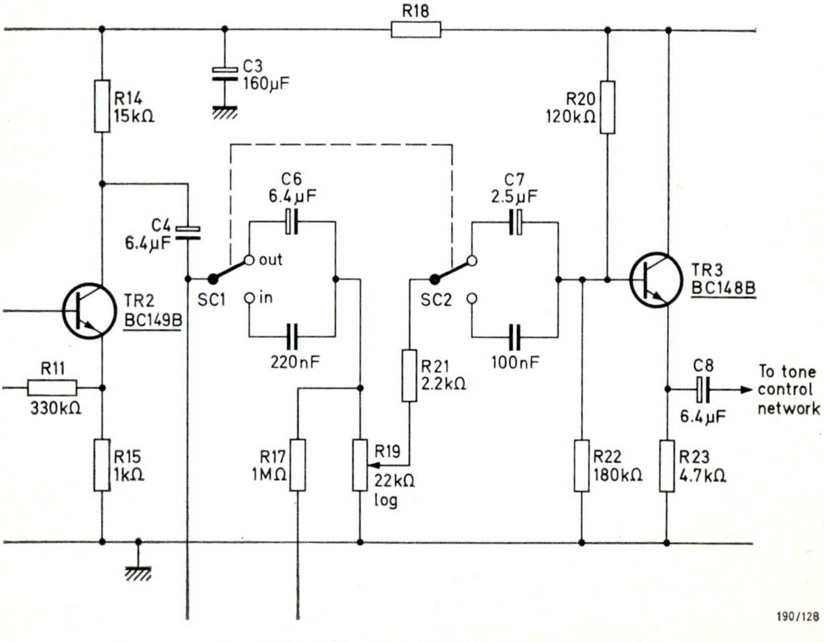

Fig. 128—Rumble filter circuit

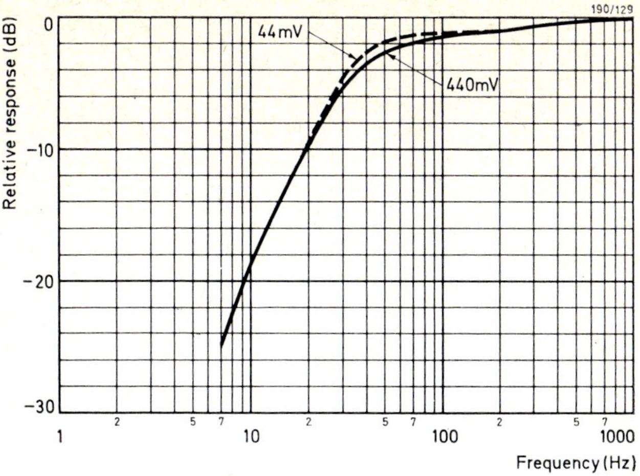

Fig. 129—Frequency characteristics of rumble filter

Scratch Filter

The circuit of the scratch filter is shown in Fig. 130. Frequency-dependent feedback is introduced across transistor TR_4, from the output of the pre-amplifier to the tone control network. Because of phase-changing effects, a peak occurs in the response characteristics before roll-off starts, and a damping resistor must be added to the RC network. To maintain the d.c. conditions, the damping resistor must be a.c. coupled to the collector of TR_4.

A parallel damping resistor sustains a sharper roll-off than a series resistor. However, the distortion of the circuit at 1kHz and 440mV output was increased from 0·04 to 0·06% (filter 'flat'), to 0·13% (filter in 5kHz position). The overload margin also dropped from 20 to 18dB. The distortion can be reduced by using a series damping resistor to 0·04% (flat) and 0·10% (5kHz), and the overload margin returned to 20dB. The filter characteristics are shown in Fig. 131 for both a parallel damping resistor of 6·8kΩ as in the circuit of Fig. 130 (full lines), and a series resistor of 4·7kΩ between the collector of TR_4 and the wiper contact of S_{D2} (broken lines).

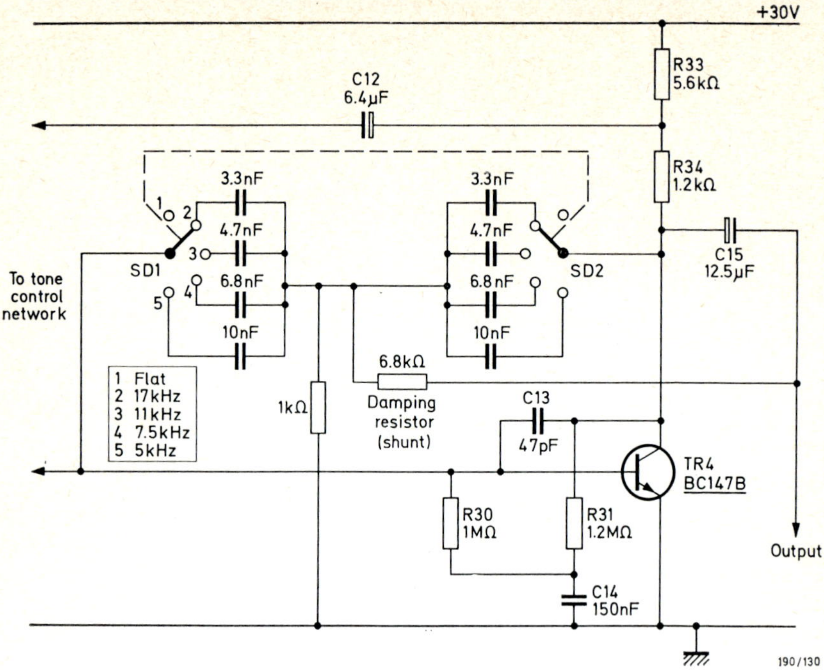

Fig. 130—Scratch filter circuit

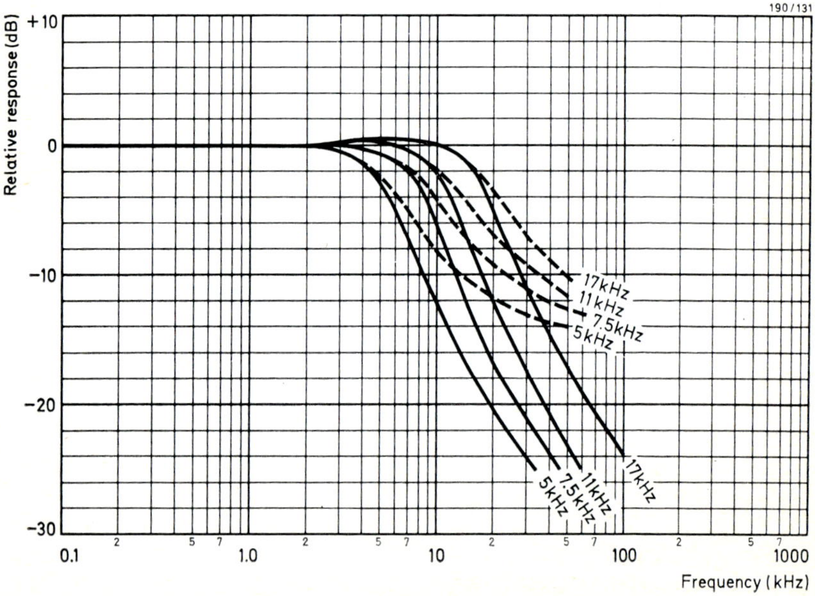

Fig. 131—Frequency characteristics of scratch filter
Full lines: parallel damping; broken lines: series damping

COMPONENT LISTS FOR CIRCUITS IN CHAPTER 7

10W Audio Amplifier (page 122)

Transistors and Rectifier

Circuit reference	Mullard type	Circuit reference	Mullard type
TR_1	BC158	TR_4 } matched pair	BD131
TR_2	BD131	TR_5 } matched pair	BD132
TR_3	BC148 or BC108	Full-wave bridge rectifier	BY164

Resistors

Mullard CR25 series (0·3W) ±5% except R_9 which should be a 2W component; R_{10}, R_{12}, and R_{13} which should be 1W components

Circuit reference	Value	Circuit reference	Value
R_1 preset potentiometer	50kΩ	R_8 preset potentiometer	100Ω
R_2	120kΩ	R_9	100Ω
R_3	100kΩ	R_{10}	270Ω
R_4	22Ω	R_{11}	10Ω
R_5	10kΩ	R_{12}	0·5Ω
R_6	470Ω	R_{13}	0·5Ω
R_7	470Ω	R_{14}	10Ω

Resistors R_2, R_3, R_5 and R_6 should be high-stability components.

Capacitors

Circuit reference	Value	Description	Mullard type
C_1	25μF	electrolytic 25V	C426AR/F25
C_2	16μF	electrolytic 40V	C426AR/G16
C_3	400μF	electrolytic 25V	C437AR/F400
C_4	10nF	polyester	C296AA/A10K
C_5	220nF	polyester	C296AA/A220K
C_6	1nF	polyester	C296AC/A1K
C_7	160μF	electrolytic 25V	C437AR/F160
C_8	47nF	polyester	C296AA/A47K
C_9	1250μF	electrolytic 25V	C431BR/F1250
C_{10}	2500μF	electrolytic 40V	C431BR/G2500

Mains Transformer

The transformer used in this circuit may be obtained from the following manufacturers under the type numbers given:

Colne Electric Limited	20043
Drake Transformers Limited	294-140
Gardners Transformers Limited	GR 97183
Parmeko Limited	P 3204

Heatsinks

The dissipation in the driver transistor may reach 2W and a 1·5mm (16 gauge) bright aluminium heatsink with an area of 10cm^2 should be used for this transistor. Each output transistor should be mounted on a 1·5mm bright aluminium heatsink with an area of 15cm^2.

15/20W Audio Amplifier (page 126)

Transistors

Circuit reference	Mullard type
TR_1	BC158
TR_2	BC147
TR_3	BC147
TR_4	BC338 or BD135
TR_5	BC338 or BD135
TR_6	BD181 } matched pair
TR_7	BD181 } matched pair

Resistors

Mullard CR25 series (0·3W) $\pm 5\%$ except R_{22} which is 6W, and R_{24} which is 12W

Circuit reference		Value	Circuit reference	Value
R_1	preset potentiometer	100kΩ	R_{13}	2·2kΩ
R_2		150kΩ	R_{14}	2·7kΩ
R_3		220kΩ	R_{15}	270Ω
R_4		68Ω	R_{16}	390Ω
R_5		56kΩ	R_{17}	100Ω
R_6		390Ω	R_{18}	330Ω
R_7		2·2kΩ	R_{19}	330Ω
R_8		270Ω	R_{20}	150Ω
R_9		10kΩ	R_{21}	150Ω
R_{10}		2·7kΩ	R_{22}	3·3Ω
R_{11}		150Ω	R_{23}	2·2Ω
R_{12}		220Ω	R_{24}	4·7Ω

Capacitors

Circuit reference	Value	Description	Mullard type
C_1	220nF	metallised film	C280AE/P220K
C_2	6·8nF	polyester	C296AC/A6K8
C_3	50μF	electrolytic 40V	C426AR/G50
C_4	250μF	electrolytic 25V	C437AR/F250
C_5	3·3nF	polyester	C296AC/A3K3
C_6	160μF	electrolytic 40V	C437AR/G160
C_7	1nF	polyester	C296AC/A1K
C_8	200μF	electrolytic 6·4V	C426AR/C200
C_9	1nF	polyester	C296AC/A1K
C_{10}	320μF	electrolytic 6·4V	C426AR/C320
C_{11}	160μF	electrolytic 25V	C437AR/F160
C_{12}	160μF	electrolytic 25V	C437AR/F160
C_{13}	2200μF	electrolytic 25V	071 16222
C_{14}	2200μF	electrolytic 40V	071 17222
C_{15}	220nF	metallised film	C280AE/P220K
C_{16}	1000μF	electrolytic 40V	071 17102
C_{17}	2200μF	electrolytic 63V	071 18222

Heatsinks

Both output transistors TR_6 and TR_7 require a heatsink of thermal resistance 3·35degC/W. (This value allows for a supply voltage 10% above nominal and may be increased if a regulated power supply is used.) Heatsinks of 3mm bright aluminium of area 175cm^2 are necessary for each output transistor. The remaining transistors do not require heatsinks.

Mains Transformer

The transformer for this circuit can be obtained from:

Colne Electric Limited	24055
Drake Transformers Limited	785-168

(see page 200)

25W Audio Amplifier (page 128)

Transistors

Circuit reference	Mullard type	
TR_1	BC158	
TR_2	BC147	
TR_3	BC148	
TR_4	BD135	matched pair
TR_5	BD136	(matched pair)
TR_6	BD201	matched pair
TR_7	BD202	(matched pair)

Resistors

Mullard CR25 series (0·3W) ±5% except R_{17} and R_{19} which are 2W components

Circuit reference	Value	Circuit reference	Value
R_1	150kΩ	R_{10}	preset potentiometer 1kΩ
R_2	150kΩ	R_{11}	680Ω
R_3	1·2kΩ	R_{12}	270Ω
R_4	47kΩ	R_{13}	270Ω
R_5	3·3kΩ	R_{14}	1·8kΩ
R_6	47Ω	R_{15}	100Ω
R_7	820Ω	R_{16}	100Ω
R_8	1·5kΩ	R_{17}	0·22Ω
R_9	1·5kΩ	R_{18}	0·22Ω
		R_{19}	10Ω

Capacitors

Circuit reference	Value	Description	Mullard type
C_1	8μF	electrolytic 40V	C426AR/G8
C_2	6·4μF	electrolytic 25V	C426AR/F6·4
C_3	160μF	electrolytic 25V	C437AR/F160
C_4	330pF	ceramic	C333CH/C330E
C_5	100pF	ceramic	C333CC/C100E
C_6	0·1μF	metallised film	C280AE/P100K
C_7	80μF	electrolytic 25V	C426AR/F80
C_8	2200μF	electrolytic 25V	071 16222

Heatsinks

The required thermal resistance of each output transistor is 8·7degC/W. Heatsinks made of 1·6mm bright aluminium for the BD201 and BD202 should be vertically mounted and measure 60cm². The dissipation of the other transistors is so low that no further heatsinks are necessary.

Protection circuit

Transistors

TR_8	BC148
TR_9	BC157

Resistors

Mullard CR25 series (0·3W) ±5%

R_{20}	8·2kΩ
R_{21}	8·2kΩ
R_{22}	preset potentiometer 4·7kΩ
R_{23}	preset potentiometer 4·7kΩ

35W Audio Amplifier (page 134)

Transistors

Circuit reference	Mullard type	
TR_1	BC157	
TR_2	BD139	
TR_3	BC148	
TR_4	BD139	matched pair
TR_5	BD140	matched pair
TR_6	BD183	matched pair
TR_7	BD183	matched pair

Resistors

Mullard CR25 series (0·3W) ±5% except R_{18} and R_{19} which are 5W components

Circuit reference	Value	Circuit reference	Value
R_1	150kΩ	R_{11}	1kΩ
R_2	150kΩ	R_{12}	220Ω
R_3	1·5kΩ	R_{13}	470Ω
R_4	33kΩ	R_{14}	470Ω
R_5	4·7kΩ	R_{15}	1·8kΩ
R_6	33Ω	R_{16}	56Ω
R_7	1kΩ	R_{17}	56Ω
R_8	2·7kΩ	R_{18}	0·5Ω
R_9	2·2kΩ	R_{19}	0·5Ω
R_{10}	preset potentiometer 1kΩ	R_{20}	10Ω

Capacitors

Circuit reference	Value	Description	Mullard type
C_1	2·5μF	electrolytic 64V	C426AR/H2·5
C_2	2·5μF	electrolytic 64V	C426AR/H2·5
C_3	10nF	metallised film	C280AE/P10K
C_4	160μF	electrolytic 40V	C437AR/G160
C_5	680pF	polystyrene	C295AC/D680E
C_6	27pF	ceramic plate	C333CC/C27E
C_7	330pF	polystyrene	425 43301
C_8	10nF	metallised film	C280AE/P10K
C_9	100nF	metallised film	C280AE/P100K
C_{10}	32μF	electrolytic 64V	C436AR/H32
C_{11}	2200μF	electrolytic 63V	071 18222

Heatsinks

The driver transistors TR_4 and TR_5 require thermal resistances of 72degC/W. Heatsinks of 1mm bright aluminium measuring 2cm^2 are necessary. The required heatsink thermal resistance for each output transistor TR_6 and TR_7 is 7·25degC/W. Bright aluminium heatsinks 64cm^2 by 2mm thick are required. No further heatsinks are required.

Protection circuit

Transistors and Diodes

TR_8	BC148
TR_9	BC157
D_1	BA148
D_2	BA148

Resistors

Mullard CR25 (0·3W) ±5% except where stated

R_{21}	10kΩ
R_{22}	10kΩ
R_{23}	preset potentiometer 4·7kΩ
R_{24}	preset potentiometer 4·7kΩ

50W Audio Amplifier (page 138)

Transistors and Diodes

Circuit reference	Mullard type	Circuit reference	Mullard type
TR_1	BC157	TR_7	BD184 } matched pair
TR_2	BC157	TR_8	BD184 } matched pair
TR_3	BD139A	D_1	BA154
TR_4	BC148	D_2	BAX16
TR_5	BD139A	D_3	BY126
TR_6	BD140A	D_4	BY126

Resistors

Mullard CR25 series (0·3W) ±5% except R_{14} which is CR37 series (0·5W) ±5%; R_{10} and R_{15} which are 1W components; and R_9, R_{19}, and R_{20} which are 2W components

Circuit reference	Value	Circuit reference	Value
R_1	2·7kΩ	R_{12}	preset potentiometer 470Ω
R_2	1kΩ	R_{13}	330Ω
R_3	18kΩ	R_{14}	3·3kΩ
R_4	56Ω	R_{15}	1kΩ
R_5	470Ω	R_{16}	56Ω
R_6	2·7kΩ	R_{17}	56Ω
R_7	12Ω	R_{18}	56Ω
R_8	1kΩ	R_{19}	0·5Ω
R_9	560Ω	R_{20}	0·5Ω
R_{10}	1·5kΩ	R_{21}	4·7Ω
R_{11}	1·5kΩ		

Capacitors

Circuit reference	Value	Description	Mullard type
C_1	12·5μF	electrolytic 25V	C426AR/F12·5
C_2	32μF	electrolytic 40V	C426AR/G32
C_3	500μF	electrolytic 2·5V	C426AR/A500
C_4	22nF	metallised film	C280AE/P22K
C_5	2·2nF	polyester	C296AC/A2K2
C_6	560pF	polystyrene	425 45601
C_7	100μF	electrolytic 40V	C437AR/G100
C_8	32μF	electrolytic 40V	C426AR/G32
C_9	100nF	metallised film	C280AE/P100K

Heatsinks

Driver transistors TR_5 and TR_6 require thermal resistances of 60degC/W. Small heatsinks of bright aluminium about 2cm^2 are adequate for these transistors.

Output transistors TR_7 and TR_8 require thermal resistances of 4degC/W. Heatsinks of 3mm bright aluminium of area 130cm^2 are needed for each output transistor. No further heatsinks are required.

Mains Transformer

The transformer for this circuit can be obtained from:

Colne Electric Limited	26078

(see page 200)

Protection Circuit

Transistors and Diodes

	Current-limit circuit	Current-trip circuit
TR_9	BC148	BC148
TR_{10}	BC158	BC158
TR_{11}	—	BC158
TR_{12}	—	BC148
D_5	OA91	—
D_6	OA91	—

Resistors

Mullard CR25 series (0·3W) ±5% except where stated

R_{22}	2·2kΩ preset potentiometer	100kΩ
R_{23}	2·2kΩ preset potentiometer	100kΩ
R_{24}	4·7kΩ	2·2kΩ preset potentiometer
R_{25}	4·7kΩ	2·2kΩ preset potentiometer
R_{26}	—	4·7kΩ
R_{27}	—	4·7kΩ

Pre-amplifier for 10W Amplifier (pages 142 and 143)

Transistors

Circuit reference	Mullard type
TR_1	BC149 or BC109
TR_2	BC148 or BC108
TR_3	BC149 or BC109

Resistors

Mullard CR25 series (0.3W) ± 5% except where stated

Circuit reference	Value	Circuit reference	Value
R_1	1·8MΩ	R_{18} logarithmic/reverse logarithmic potentiometer	10kΩ
R_2	see text		
R_3	56kΩ	R_{19}	22kΩ
R_4	220kΩ	R_{20} linear potentiometer	250kΩ
R_5	22kΩ	R_{21} linear potentiometer	100kΩ
R_6	100kΩ	R_{22}	22kΩ
R_7	100kΩ	R_{23}	15kΩ
R_8	12kΩ	R_{24}	47kΩ
R_9	220kΩ	R_{25}	1kΩ
R_{10}	680Ω	R_{26}	47kΩ
R_{11}	560Ω	R_{27}	220kΩ
R_{12}	6·8kΩ	R_{28}	10kΩ
R_{13}	12kΩ	R_{29}	15kΩ
R_{14}	22kΩ	R_{30}	220kΩ
R_{15}	560kΩ	R_{31}	3·3kΩ
R_{16}	10kΩ	R_{32}	6·8kΩ
R_{17} logarithmic potentiometer	10kΩ	R_{33}	3·9kΩ

Resistors R_6, R_7, R_8 and R_9 should be high-stability components.

Capacitors

Circuit reference	Value	Description	Mullard type
C_1	6·4μF	electrolytic 25V	C426AR/F6·4
C_2	80μF	electrolytic 2·5V	C426AR/A80
C_3	2·5μF	electrolytic 25V	C426AR/H2·5
C_4	10nF	metallised foil	C281AB/A10K
C_5	3·3nF	polyester	C296AC/A3K3
C_6	10nF	metallised foil	C281AB/A10K
C_7	10μF	electrolytic 16V	C426AR/E10
C_8	2·5μF	electrolytic 25V	C426AR/H2·5
C_9	80μF	electrolytic 25V	C426AR/F80
C_{10}	15nF	metallised foil	C281AB/A15K
C_{11}	15nF	metallised foil	C281AB/A15K
C_{12}	10μF	electrolytic 16V	C426AR/E10
C_{13}	3·3nF	polyester	C296AC/A3K3

Circuit reference	Value	Description	Mullard type
C_{14}	3·3nF	polyester	C296AC/A3K3
C_{15}	10μF	electrolytic 16V	C426AR/E10
C_{16}	6·8nF	polyester	C296AC/A6K8
C_{17}	470nF	metallised foil	C281AB/A470K
C_{18}	10μF	electrolytic 16V	C426AR/E10
C_{19}	470nF	metallised foil	C281AB/A470K
C_{20}	6·8nF	polyester	C296AC/A6K8
C_{21}	2·2nF	polyester	C296AC/A2K2
C_{22}	10μF	electrolytic 16V	C426AR/E10

Universal Pre-amplifier (pages 148 and 149)

Resistors

Mullard CR25 series (0·3W) ±5% except where stated. Resistors R_9, R_{11}, and R_{30} should be low-noise types

Circuit reference	Value	Circuit reference	Value	
R_1	1MΩ	R_{19}	22kΩ	logarithmic potentiometer
R_2	39kΩ			
R_3	470kΩ	R_{20}	120kΩ	
R_4	39kΩ	R_{21}	2·2kΩ	
R_5	470kΩ	R_{22}	180kΩ	
R_6	39kΩ	R_{23}	4·7kΩ	
R_7	39kΩ	R_{24}	6·8kΩ	
R_8	56kΩ	R_{25}	100kΩ	linear potentiometer
R_9	220kΩ	R_{26}	12kΩ	
R_{10}	680Ω	R_{27}	2·2kΩ	
R_{11}	330kΩ	R_{28}	4·7kΩ	linear potentiometer
R_{12}	22kΩ	R_{29}	6·8kΩ	
R_{13}	82kΩ	R_{30}	1MΩ	
R_{14}	15kΩ	R_{31}	1·2MΩ	(±10%)
R_{15}	1kΩ	R_{32}	22kΩ	linear potentiometer
R_{16}	68kΩ	R_{33}	5·6kΩ	
R_{17}	1MΩ	R_{34}	1·2kΩ	
R_{18}	4·7kΩ mono	R_{35}	220Ω	
	2·7kΩ stereo	R_{36}	820Ω	

Transistors

Circuit reference	Mullard type
TR_1	BC149C
TR_2	BC149B
TR_3	BC148B
TR_4	BC147B

Capacitors

Circuit reference	Value	Description	Mullard type
C_1	1·6μF	electrolytic 25V	C426AS/F1·6
C_2	4·7nF	polyester	C296AC/A4K7
C_3	160μF	electrolytic 25V	C437AR/F160
C_4	6·4μF	electrolytic 25V	C426AR/F6·4
C_5	1·5nF	polyester	C296AC/A1K5
C_6	6·4μF	electrolytic 25V	C426AR/F6·4
C_7	2·5μF	electrolytic 16V	C426AS/E2·5
C_8	6·4μF	electrolytic 25V	C426AR/F6·4
C_9	47nF	metallised film	C280AE/P47K
C_{10}	4·7nF	polyester	C296AC/A4K7
C_{11}	47nF	metallised film	C280AE/P47K
C_{12}	6·4μF	electrolytic 25V	C426AR/F6·4
C_{13}	47pF	ceramic plate	C333CC/C47E
C_{14}	150nF	metallised film	C280AE/P150K
C_{15}	12·5μF	electrolytic 25V	C426AR/F12·5
C_{16}	6·4μF	electrolytic 25V	C426AR/F6·4

Mains Transformer

The transformer used in this circuit can be obtained from:

Colne Electric Limited	24056
Drake Transformers Limited	181-310

(see page 200)

Power Supplies for Amplifiers and Pre-amplifiers

The power supplies for the 15/20W, 25W, 35W, and 50W amplifiers have the same configuration as that for the 10W amplifier on page 123. The output voltage of the transformer is chosen to give the required supply voltage after rectification. The transformer secondary voltage and current for each amplifier is given below. Where a transformer is available from a manufacturer, the type number is given in the components list for the amplifier.

Amplifier	Transformer r.m.s. secondary voltage (V)	Transformer current (A)
15/20W	34	2·2
25W	27	4·0
35W	53	3·4
50W	25-0-25	3·5

All power supplies use 4 × BYX22 diodes in the rectifier bridge.

The power supply for the universal pre-amplifier uses a transformer with an r.m.s. secondary voltage of 34V and a BY164 bridge rectifier. The supply requires more smoothing than those for the amplifiers, and the single electrolytic capacitor is replaced by an RC filter. This consists of a 680μF capacitor (nearest the rectifier) and a 400μF capacitor. The resistor is chosen to give a voltage of 30V at the input supply of the pre-amplifier.

CHAPTER 8

HIGH-QUALITY F.M. TUNERS

Two f.m. tuners are described, the first using a BF200 silicon planar transistor in the r.f. stage and the second using a BFW10 field-effect transistor. The tuners may be used with the f.m. stages of the i.f. amplifier described in Chapter 4.

A comparison of the two tuners will reveal that the circuits give very similar performances. However, it should be noted that the FET tuner achieves this performance at the expense of a relatively high supply voltage and high operating current in the r.f. stage. In addition, the field-effect transistor must be selected for a value of I_{DSS} (drain current at zero source-gate voltage) of 8 to 20mA.

The chapter also includes a stereo decoder circuit for reception of stereophonic broadcasts.

The difficulties associated with the construction and lining-up of r.f. and i.f. amplifiers must again be stressed. Only those constructors with previous experience in this type of circuitry and with access to specialised equipment should consider building the circuits described in this chapter.

Component lists for the circuits are given at the end of the chapter.

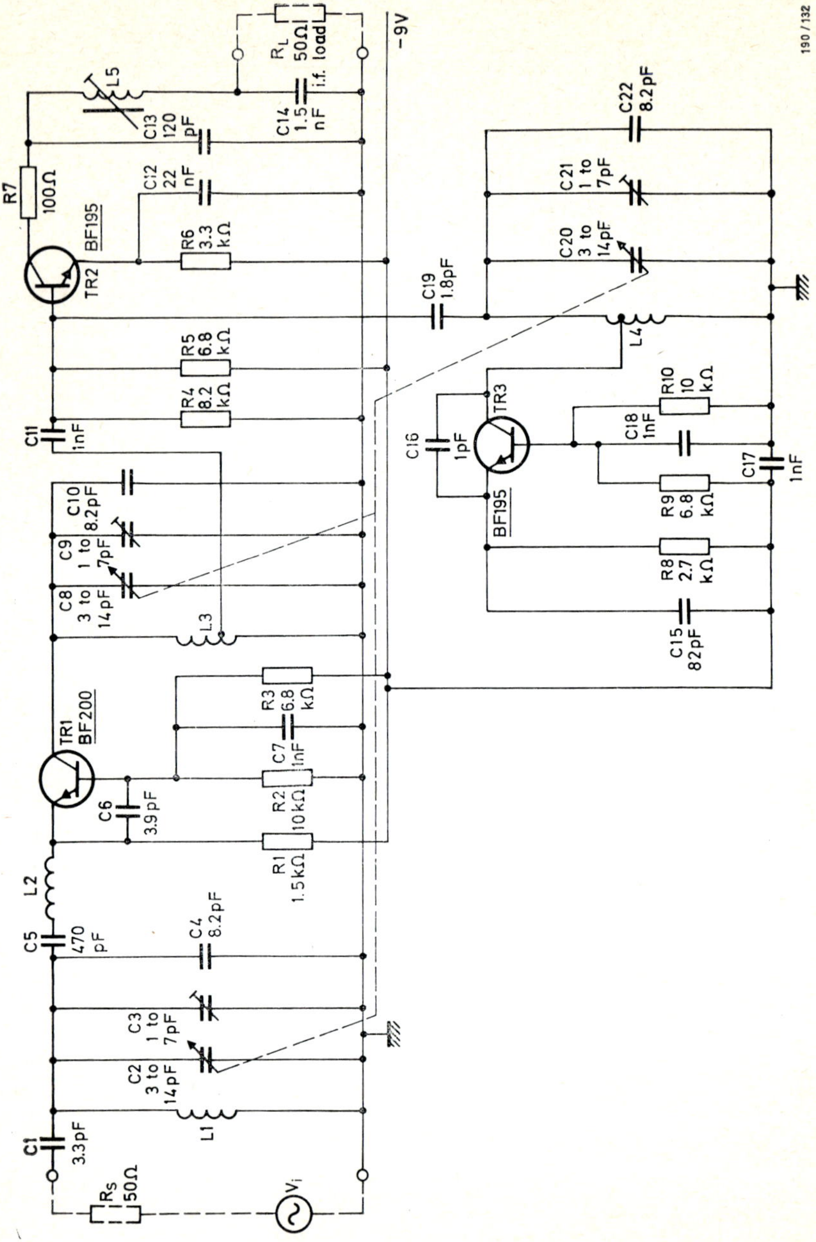

Fig. 132—Three-ganged, capacitively-tuned f.m. tuner circuit

F.M. TUNER USING BF200 SILICON PLANAR TRANSISTOR

High-quality f.m. reception calls for a tuner having a low noise factor and excellent signal-handling ability. These objectives can be achieved by exploiting the inherently good signal-handling capability of the BF200 in a circuit configuration favouring optimum noise performance.

This high-performance tuner uses a tuned aerial circuit matched to the r.f. transistor by a fixed inductance. By this means it is possible to retain the benefits of a low-loss selective aerial circuit, which can be approximately power matched, and, through an impedance transform, to allow the transistor to be matched for optimum noise performance. The transistor is then required to handle less signal for a given noise performance.

Circuit Description

The complete circuit of a three-ganged capacitively-tuned f.m. tuner is given in Fig. 132.

R.F. Stage

The BF200 transistor, TR_1, operates in the common-base configuration at a collector current of 2mA. The optimum source impedance for minimum noise is equivalent to the parallel combination of 125Ω resistance and 500Ω inductance, giving a typical noise factor of 2dB.

The emitter resistor R_1, inductance L_2 and capacitor C_6 cause some additional losses which result in a typical noise factor for the tuner of about 4·5dB. Capacitor C_6 prevents parasitic oscillation at very high frequencies and must be mounted between the base and emitter leads of the BF200 and as closely as possible to them.

The low feedback capacitance of the BF200 allows the collector to be connected to the top of the tuned interstage circuit. The collector output damping of the BF200 is negligible.

The transducer gain of the r.f. stage is about 16dB at a 3dB bandwidth of the loaded interstage circuit of 1MHz. The value of Q_O of this tuned circuit is 200.

Mixer Stage

The mixer transistor, TR_2, is a Lockfit transistor, type BF195, and operates in the common-emitter configuration at a collector current of 1mA. At this current and with a source impedance of about 100Ω, the contribution of the mixer stage to the total noise factor is small—approximately 0·5dB.

By keeping the oscillator signal at the base of TR_2 to as low a value as possible—about 50mV—harmonics of the oscillator frequency at the input to the mixer are virtually eliminated, while sufficient conversion gain is maintained.

Because the base of TR_2 has an inductive tap on the interstage circuit, no additional 10·7MHz i.f. series filter is necessary between the base and the emitter. To prevent i.f. parasitic oscillations, a 100Ω resistor, R_7, is connected in series with the collector.

The value of Q_O of the i.f. tuned circuit is about 150 and the value of Q_L is 50. The mixer gain is about 18dB.

Oscillator Stage

The oscillator transistor, TR_3, is also a BF195, and operates in the common-base configuration at a current of 1mA. The coupling between the top of the oscillator coil and the base of the mixer is obtained via capacitor C_{19}. The low value of this capacitance—1·8pF—guarantees that large aerial signals have practically no effect on the oscillator circuit. The oscillator frequency is below the aerial frequency.

Performance

In obtaining the performance data given in this section, the aerial and i.f. trimmer capacitors were adjusted at each frequency to give no padding error at the measuring point.

Transducer Gain

Transducer gain is defined as the ratio of the power delivered to the load to the power supplied into the stage if it were power-matched to the source, that is if the input impedance were equal to the source impedance.

$$\text{Transducer gain} = \frac{V_O{}^2}{R_L} \cdot \frac{4R_S}{V_S{}^2}$$

where V_O = voltage developed across load
R_L = load impedance
R_S = source impedance of generator
V_S = e.m.f. of generator

The values of transducer gain at three tuning frequencies are given in Table 5. A curve of transducer gain at a tuning frequency of 98MHz plotted against supply voltage is given in Fig. 133.

TABLE 5

Measured Values of Transducer Gain

Tuning Frequency (MHz)	Transducer Gain (dB)
90	32
98	35
105	35

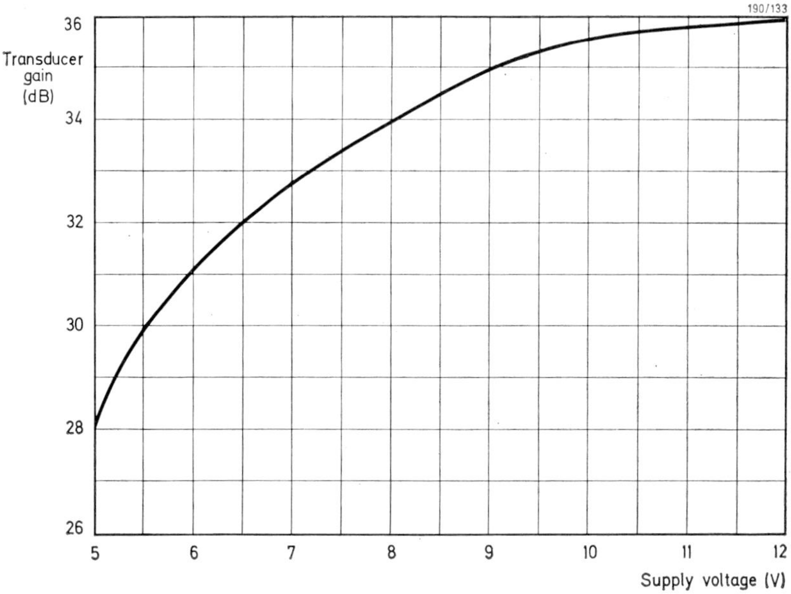

Fig. 133—Variation of transducer gain with supply voltage at a tuning frequency of 98MHz

Noise Factor

The values of noise factor at three tuning frequencies are given in Table 6.

TABLE 6

Measured Values of Noise Factor

Tuning Frequency (MHz)	Noise Factor (dB)
90	4·7
98	4·5
105	4·3

Repeat-spot Suppression

Repeat spots are caused by harmonics of the oscillator signal mixing with those of a strong aerial signal resulting in a signal having the intermediate frequency. For example, if an unwanted aerial signal of 88MHz is present, its second harmonic (176MHz) mixes with the second harmonic (165·3MHz) of an oscillator signal of 82·65MHz to give 10·7MHz—the intermediate frequency. The 82·65MHz oscillator signal corresponds to an aerial signal of 93·35MHz and so the 88MHz aerial signal will be received when the dial is set to 93·35MHz.

Repeat-spot suppression is sometimes known as half i.f. rejection. The values of repeat-spot suppression are given for two input levels at each of three tuning frequencies in Table 7. The value of R_S is 50Ω throughout. A curve of repeat-spot suppression against input voltage at a tuning frequency of 98MHz is given in Fig. 134. From this curve it can be seen that the repeat-spot suppression improves by 1dB for a decrease in signal of 2dB.

TABLE 7
Measured Values of Repeat-spot Suppression

Tuning Frequency (MHz)	Input Voltage (μV)	Repeat-spot Suppression (dB)
90	40	73
90	1	89
98	40	71
98	1	87
105	40	70
105	1	86

Double-beat Suppression

Double beats occur if two strong aerial signals have frequencies or harmonics which combine with the oscillator signal, or its harmonics, to give the intermediate frequency. For example, if two strong unwanted aerial signals with frequencies of 94MHz and 96·7MHz are present, and the dial is set to 100·7MHz, the audio signals associated with the two strong signals will be heard superimposed upon each other. This is because the dial setting of 100·7MHz corresponds to an oscillator frequency of 90MHz and the second harmonic of this oscillator signal (180MHz) combines with the two strong unwanted aerial signals (94+96·7−180) to give 10·7MHz—the intermediate frequency.

The values of double-beat suppression given in Table 8 were measured with two identical interference signals at 95 and 97·5MHz. Double-beat reception is possible at the tuning frequencies 92·5, 100 and 101·6MHz and at these frequencies, measurements of double-beat suppression for a required signal level of 40μV are given. The value of R_s is 50Ω. A curve of double-beat suppression against input voltage is given in Fig. 135.

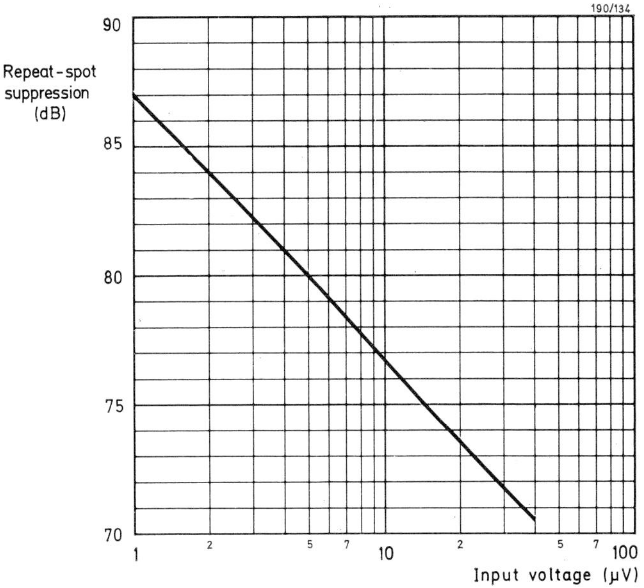

Fig. 134—Variation of repeat-spot suppression with input voltage at a tuning frequency of 98MHz

TABLE 8

Measured Values of Double-beat Suppression

Tuning Frequency (MHz)	Double-beat Suppression (dB)
92·5	63
100	66
101·6	65

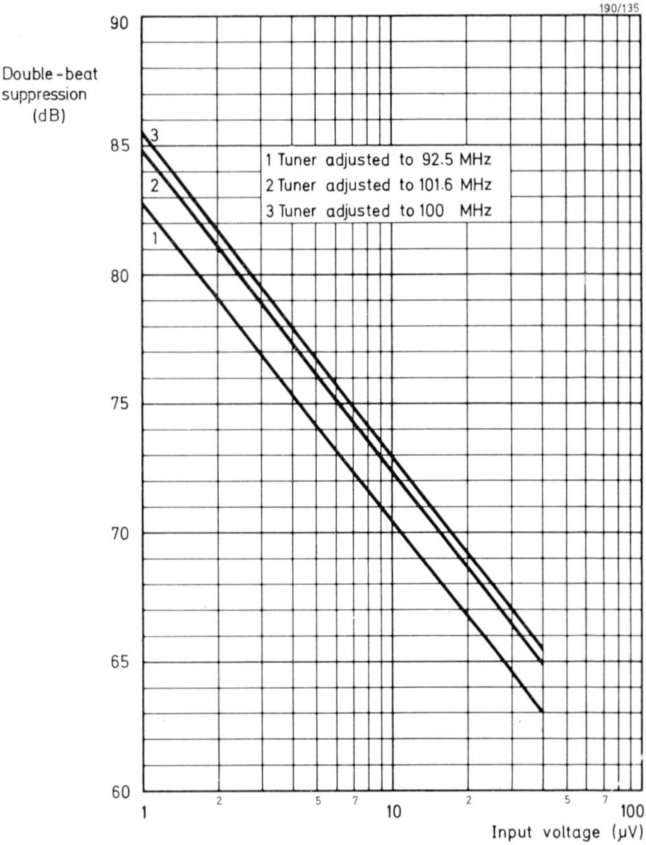

Fig. 135—Variation of double-beat suppression with input voltage

Continuous-beat Suppression

Continuous-beats occur when two strong aerial signals have frequencies, or harmonics, which combine to give the intermediate frequency. If continuous beats occur, they occur regardless of the dial setting. For example, if two strong aerial signals are present with frequencies of 90MHz and 95·35MHz, their second harmonics (180MHz and 190·7MHz respectively) combine to produce 10·7MHz—the intermediate frequency.

The continuous-beat suppression, measured at a tuning frequency of 98MHz, is 67dB at input voltage levels of 40μV for both signals.

Spurious-response Suppression

The one-signal spurious-response suppression is given in histogram form in Fig. 136. The tuner frequency is 98MHz and the reference level of the required input signal is 8μV. The spikes in Fig. 136 show that at a certain input frequency, an i.f. output signal occurs. The input signal level at this frequency must be many times higher to obtain the same i.f. output signal as for a wanted input signal of 8μV. The ratio of the required interference input signal level and the required input signal of 8μV is given as spurious-response suppression. This ratio is indicated by the top level of the spikes in Fig. 136.

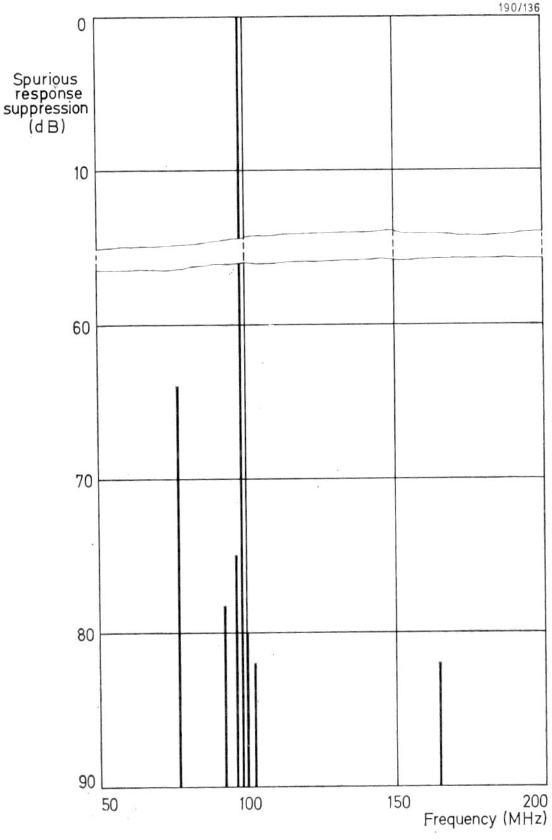

Fig. 136—Histogram of spurious-response suppression

Image Rejection

The image rejection is determined by the selectivity of the aerial at a frequency different from the tuned frequency, and by interstage circuit quality factors. In the tuner described, the image-rejection frequency is $2 \times 10{\cdot}7$MHz below the tuning frequency. The values of image rejection at three tuning frequencies are given in Table 9.

TABLE 9
Measured Values of Image Rejection

Tuning Frequency (MHz)	Image Rejection (dB)
90	63
98	64
105	63

I.F. Rejection

The values of i.f. rejection at three tuning frequencies are given in Table 10.

TABLE 10
Measured Values of I.F. Rejection

Tuning Frequency (MHz)	I.F. Rejection (dB)
90	64
98	63
105	61

Signal Handling

When the value of R_S is 50Ω, signals of more than 1V can be handled. The shift of the oscillator frequency is not more than 20kHz for signals up to 1V.

Oscillator Voltage on Mixer

The oscillator voltage at the base of the mixer is kept low. The values of oscillator voltage at three tuning frequencies are given in Table 11.

Frequency Shift of Oscillator

The oscillator frequency shift due to supply voltage variations is about 20kHz per 2V supply voltage variation.

TABLE 11
Measured Values of Oscillator Voltage at Base of Mixer

Tuning Frequency (MHz)	Oscillator Voltage (mV)
90	35
98	50
105	60

Oscillator Radiation

When the value of R_S is 50Ω, the oscillator radiation at the input terminals of the tuner is 300μV.

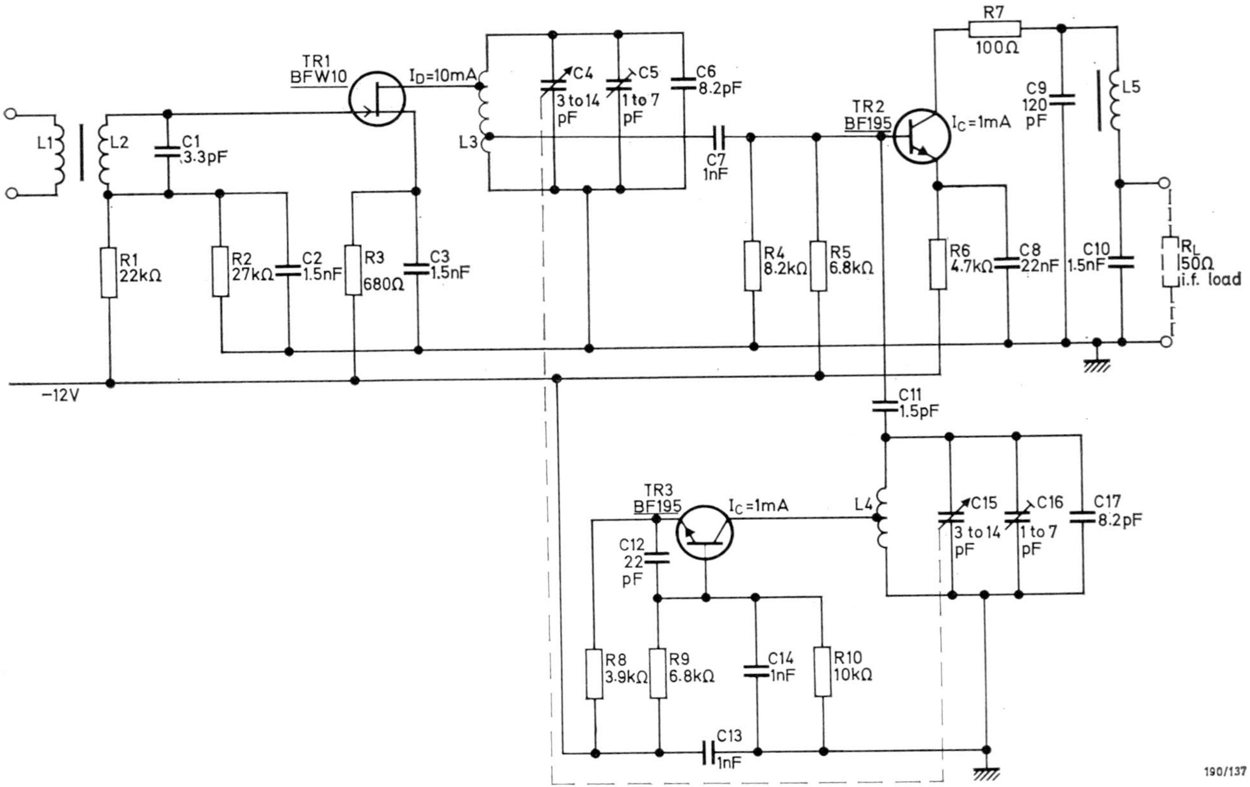

Fig. 137—Two-ganged, capacitively-tuned f.m. tuner circuit

F.M. TUNER USING BFW10 FIELD-EFFECT TRANSISTOR

The circuit for a high-quality, two-ganged, capacitively-tuned f.m. tuner is shown in Fig. 137. The tuner uses a BFW10 field-effect transistor in the r.f. input stage, and BF195 transistors in the mixer and oscillator stages.

Circuit Description

R.F. Stage

The BFW10 field-effect transistor operates at a typical drain current of 10mA and typical drain-source voltage of 6V. The optimum source impedance for minimum noise is equivalent to the parallel combination of 910Ω resistance and 370Ω inductance, giving a typical noise factor of 2·5dB.

The aerial input circuit is transformer-coupled to the input of the field-effect transistor, the unloaded quality factor, Q_O, of the aerial transformer being about 80 at a secondary inductance of 330nH.

The transducer gain of the r.f. stage is about 8·5dB at 100MHz. The value of Q_O for the interstage circuit is 200.

The d.c. stabilisation is obtained by a voltage divider for the gate electrode and a source resistance of value 680Ω. A limitation of this design is that the minimum value of I_{DSS} for the field-effect transistor is 8mA because the gate of the transistor may not be positive with respect to the source.

Mixer and Oscillator Stages

Comparison of the circuit diagram of this tuner with that of the planar-transistor version shown on page 202 will show that the mixer and oscillator stages of the two circuits are almost identical. The description given on pages 203 and 204 applies equally to this tuner.

Performance

For definitions of terms transducer gain, repeat-spot suppression, double-beat suppression, continuous-beat suppression, spurious-response suppression and image rejection, see pages 204 to 210.

Transducer Gain

The values of transducer gain measured at three tuning frequencies are given in Table 12. In Fig. 138 the transducer gain at a tuning frequency of 98MHz is plotted as a function of supply voltage.

TABLE 12
Measured Values of Transducer Gain

Tuning Frequency (MHz)	Transducer Gain (dB)
90	26
98	30
106	29

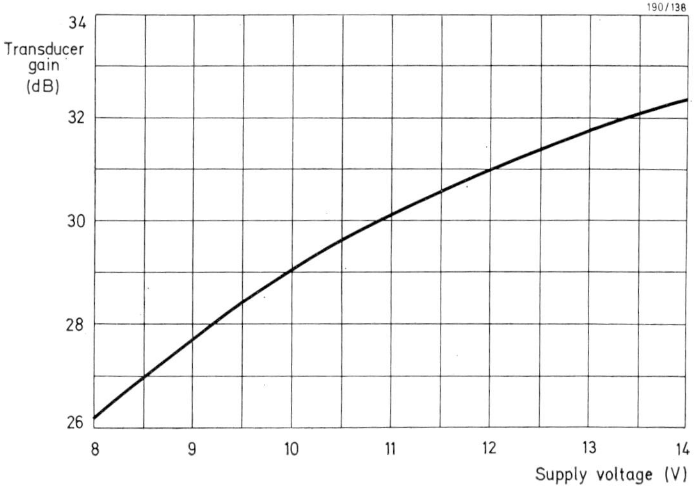

Fig. 138—Variation of transducer gain with supply voltage at a tuning frequency of 98MHz

Noise Factor

The values of noise factor measured at three tuning frequencies are given in Table 13.

In this circuit, the noise contribution of the mixer stage rises when the power gain of the r.f. stage decreases. Because of the selectivity of the aerial input circuit, the power gain of the r.f. stage at 90MHz is 3dB below that at 98MHz, hence the increase in noise factor at the lower frequency.

Repeat-spot Suppression

The values of repeat-spot suppression at three tuning frequencies and an aerial signal level of 100μV are given in Table 14.

TABLE 13
Measured Values of Noise Factor

Tuning Frequency (MHz)	Noise Factor (dB)
90	6·9
98	4·2
106	4·1

TABLE 14
Measured Values of Repeat-spot Suppression

Tuning Frequency (MHz)	Repeat-spot Suppression (dB)
90	68
98	61
106	58

For an aerial signal level of 1μV the repeat-spot suppression would be about 80dB.

Double-beat Suppression

The values of double-beat suppression at three tuning frequencies are given in Table 15.

TABLE 15
Measured Value of Double-beat Suppression

Tuning Frequency (MHz)	Double-beat Suppression (dB)
92·5	60
100	57
101·6	59

Continuous-beat Suppression

The values of continuous-beat suppression at three tuning frequencies with both aerial signals at a level of 100μV are given in Table 16.

TABLE 16

Measured Values of Continuous-beat Suppression

Tuning Frequency (MHz)	Continuous-beat Suppression (dB)
90	65
98	60
106	56

Spurious-response Suppression

The one-signal spurious-response suppression is given in histogram form in Fig. 139. The tuning frequency is 98MHz. The reference level of the required input signal is 20μV.

Image Rejection

The values of image rejection at three tuning frequencies are given in Table 17.

TABLE 17

Measured Values of Image Rejection

Tuning Frequency (MHz)	Image Rejection (dB)
90	42
98	43
106	38

I.F. Rejection

The values of i.f. rejection at three tuning frequencies are given in Table 18.

TABLE 18

Measured Value of I.F. Rejection

Tuning Frequency (MHz)	I.F. Rejection (dB)
90	76
98	76
106	75

Signal Handling

With a value of source impedance of 300Ω, signals of more than 1V can be handled. The shift of oscillator frequency is not more than 20kHz for signals up to 1V.

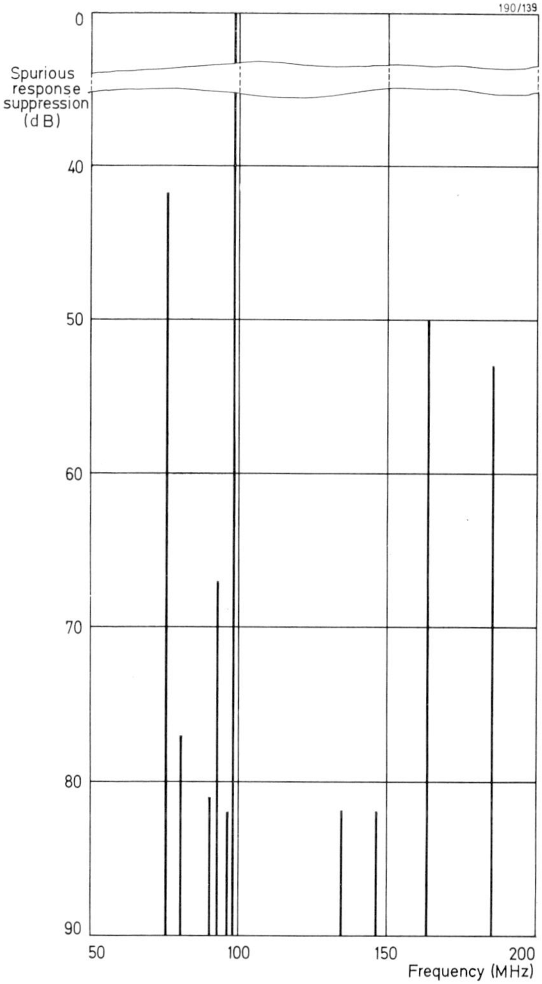

Fig. 139—Histogram of spurious-response suppression

Oscillator Voltage on Mixer

The oscillator voltages at the base of the mixer at three tuning frequencies are given in Table 19.

TABLE 19
Oscillator Voltage at Base of Mixer

Tuning Frequency (MHz)	Oscillator Voltage (mV)
90	40
98	50
106	80

Frequency Shift of Oscillator

The frequency shift of the oscillator due to a supply voltage variation of 1V is less than 20kHz. The lower limit of supply voltage for oscillator operation is 4V.

STEREO DECODER

Stereo Transmissions

Stereophonic programmes are regularly broadcast on the v.h.f. transmissions of BBC Radio Three. For these transmissions to be received stereophonically, certain equipment in addition to the normal v.h.f. receiver is required. This additional equipment consists of two separate audio amplifiers and loudspeakers to amplify the left- and right-hand channels and some means of separating—or decoding—the channels.

The system the BBC, in common with other European broadcasting authorities, uses is the pilot-tone (Zenith-G.E.) system, with a 50μs pre-emphasis time-constant instead of the 75μs used in the U.S.A. In this system, the left- and right-hand channels are encoded at the transmitter by a multiplexing process and transmitted on a single wavelength. The channels are decoded after detection in the receiver. The system is compatible; that is, listeners with unadapted v.h.f. receivers can receive an acceptable monophonic representation of the stereophonic broadcast.

The sum of the left- and right-hand channels provides a satisfactory monophonic signal which can be transmitted and received in the normal way. For stereophonic reception, however, additional information is necessary to enable the stereo receiver to separate the left- and right-hand signals. In the Zenith-G.E. system, this extra information consists of a 'difference' signal—left minus right—which is transmitted as the upper and lower sidebands of a 38kHz subcarrier. The 38kHz subcarrier is itself totally suppressed.

To ensure synchronisation between the decoder in the receiver and the encoder at the transmitter, a low-level synchronising signal, or pilot tone, is also transmitted. The frequency of the pilot tone is 19kHz.

Circuit Description

The circuit diagram of a decoder is shown in Fig. 140. The pilot tone is extracted from the multiplex input by means of a 19kHz circuit immediately following the discriminator output. The pilot tone is then amplified by the first transistor and applied to the base of the second transistor. The second transistor is biased to operate in class B; that is, only the positive half-cycles of the amplified pilot tone are amplified. The distorted waveform resulting from this type of amplification is very rich in second harmonic content, and this second harmonic (38kHz) is extracted by means of a tuned circuit at the collector of the second transistor. The peak-to-peak amplitude of the 38kHz signal, between the collector end of the coil and the tap on the coil, is limited to approximately twice the supply voltage; hence its amplitude is constant regardless of input level once a certain value has been exceeded.

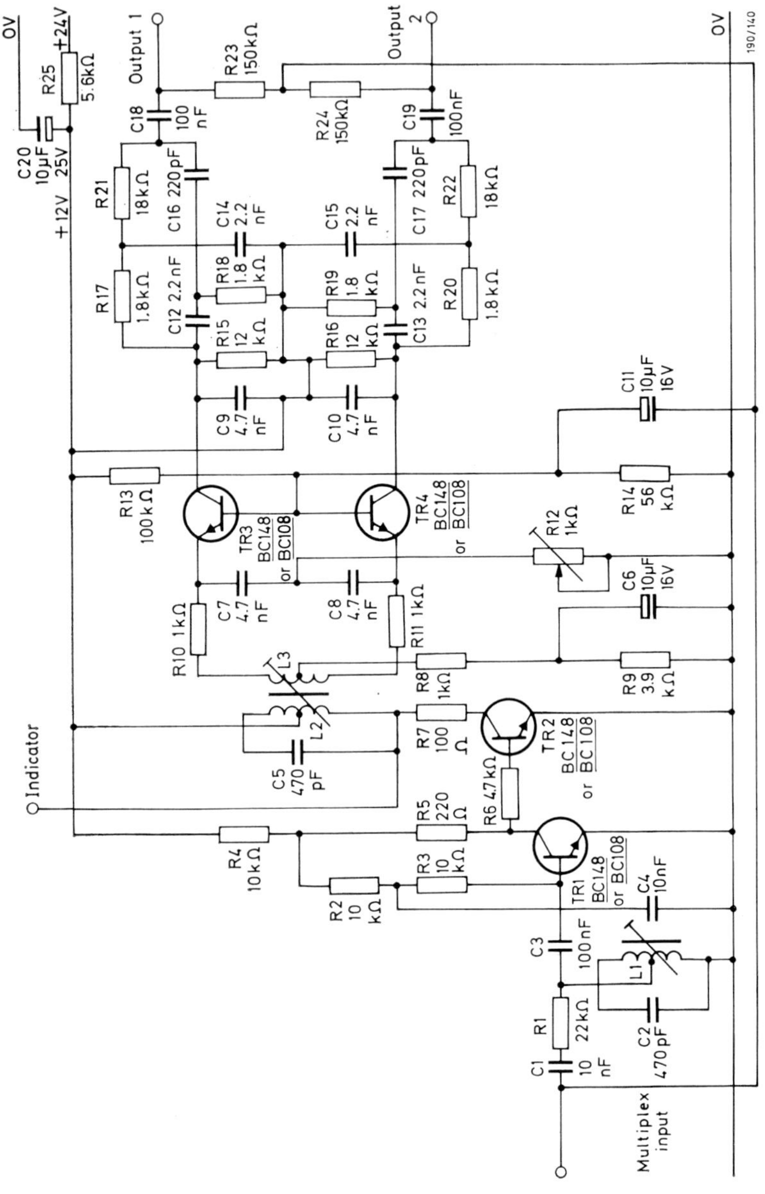

Fig. 140—Stereo decoder circuit

A centre-tapped secondary winding applies antiphase switching signals to the emitters of two transistors and the complete multiplex information is applied to their bases. With this arrangement, synchronous detection to obtain the 'difference' signal, along with 'matrixing' of the 'sum' and 'difference' signals, occurs and the two separated left- and right-hand signals appear—one at each collector. With a simple decoder operating on the principle described, only a limited separation is obtainable, so arrangements have been made in the design of the circuit shown in Fig. 140 to improve the separation. This is done by deliberately introducing a controlled amount of crosstalk into the two outputs in antiphase to the inherent crosstalk, thereby cancelling the crosstalk.

Conventional de-emphasis occurs in the collector circuits of the two switching transistors but this is insufficient to remove the 38kHz waveform from the output. As the 38kHz waveform is of sufficient amplitude to constitute a nuisance, in that it may give rise to distortion in any following audio amplifier or to beating with the oscillator frequency of a tape recorder, it is removed by passing each output through a parallel-T filter tuned to 38kHz. This does have a slight adverse effect on the de-emphasis but it is kept within acceptable limits.

In the absence of the 38kHz switching waveform at the emitters of the two switching transistors, they are biased to operate in class A and any signal applied to their bases is reproduced at their collectors. The circuit is therefore compatible, requiring no switching between mono and stereo operation.

The input impedance of the decoder is approximately 15kΩ and the decoder may be fed directly from a semiconductor discriminator capable of providing a loaded output with a peak-to-peak amplitude of approximately 750mV. It may also be fed from a typical valve discriminator, in which case the discriminator output should be attenuated by means of a series resistor shunted by a capacitor. Typical values for the resistor and capacitor are 3·3kΩ to 33kΩ and a few hundred picofarads. Such an arrangement not only provides the correct input level at all multiplex frequencies but also reduces the discriminator loading to an acceptable value.

Each of the audio amplifiers into which the outputs of the decoder are fed must have an input impedance of not less than 50kΩ.

Stereo Indicator

A stereo indicator gives a visual indication when a stereo transmission is in progress. It is also essential in the alignment of the decoder if only a limited amount of test equipment is available. A stereo indicator circuit using a single transistor is shown in Fig. 141.

This stereo indicator circuit operates by detecting the 38kHz signal generated when the 19kHz pilot tone is passed through the frequency-doubling circuit in the decoder. This 38kHz signal is fed through a limiting resistor and blocking capacitor to a shunt rectifier. The voltage derived from this shunt rectifier is applied to the base of a transistor. The transistor remains cut off when the signal is absent and is turned on when it is present. Therefore, the bulb in the collector circuit glows when a stereophonic broadcast is in progress but remains extinguished when a monophonic transmission is being received.

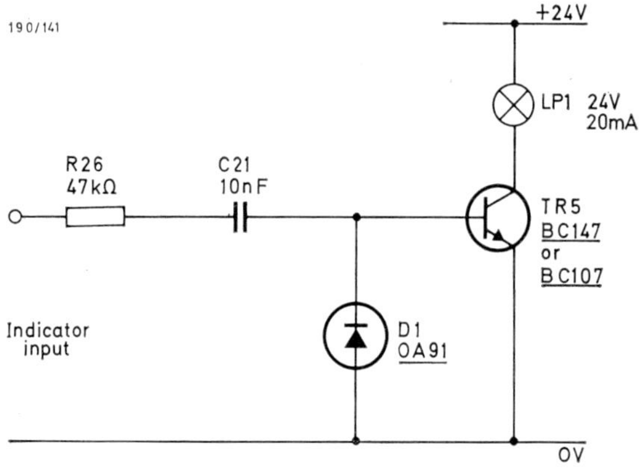

Fig. 141—Stereo indicator circuit

Alignment of the Decoder

To assist those people who do not have access to a wide range of accurate test equipment, a simple alignment procedure using the BBC transmissions has been devised. Tones are transmitted by the BBC outside normal listening hours to enable the channels to be identified and to assist in alignment. Details of the broadcasts are given in Appendix 2. For the initial alignment, however, any stereo broadcast can be used. The alignment procedure is as follows.

First, remove the de-emphasis component network from the discriminator output.

Connect a potentiometer with a value of at least 50kΩ across the discriminator output and connect the decoder input to the potentiometer slider. With the potentiometer set for maximum output, tune the receiver to a stereo broadcast on Radio Three.

Adjust the 19kHz and 38kHz coil slugs to give maximum output at the 'Indicator' terminal (maximum brightness of the indicator lamp). Adjust the input potentiometer setting until the indicator lamp just glows, then adjust the coil slugs again to give maximum brightness. Keep repeating this procedure until adjusting the cores either way results in the lamp getting dimmer. The two coils are now aligned. If tolerances in coil manufacture result in difficulty being experienced in this adjustment, especially of the 38kHz coil, because the slugs are at one end of their travel, replace the 470pF capacitor across the coil by a 390pF capacitor.

When the coils are aligned, connect the discriminator output directly to the decoder input and discard the input potentiometer.

Adjust the preset potentiometer in the decoder circuit to its mid-position. The decoder should give a reasonable degree of separation, but further adjustment using the test-tone transmissions is necessary, as follows.

Connect the two outputs to two audio amplifiers and identify the left- and right-hand channels. When the left-hand channel only is being transmitted, disconnect the left-hand amplifier and turn the 38kHz slug anti-clockwise (when viewed from the top) until the output from the right-hand loudspeaker is a minimum. Anti-clockwise rotation is necessary because a false minimum is encountered if the slug is rotated clockwise.

Reconnect the left-hand amplifier.

When the right-hand channel only is being transmitted, disconnect the right-hand amplifier and adjust the preset potentiometer from its mid-point position to give minimum output from the left-hand loudspeaker.

Reconnect the right-hand amplifier.

The decoder is now correctly aligned.

Performance

The channel separation of the decoder is shown graphically in Fig. 142. The signal-to-noise ratio is well above the acceptable level although the signal-to-noise ratio of a stereo broadcast is inherently slightly lower than that of a mono broadcast. For good stereo reproduction, the i.f. bandwidth of the tuner unit should be not less than 180kHz.

Aerial Requirements for Stereo Reception

One common cause of poor results from an f.m. stereo receiver is an inadequate aerial. Satisfactory stereo results cannot be expected unless a really good signal is fed into the tuner.

Unlike the other signals in the complex f.m. stereo signal, the 19kHz

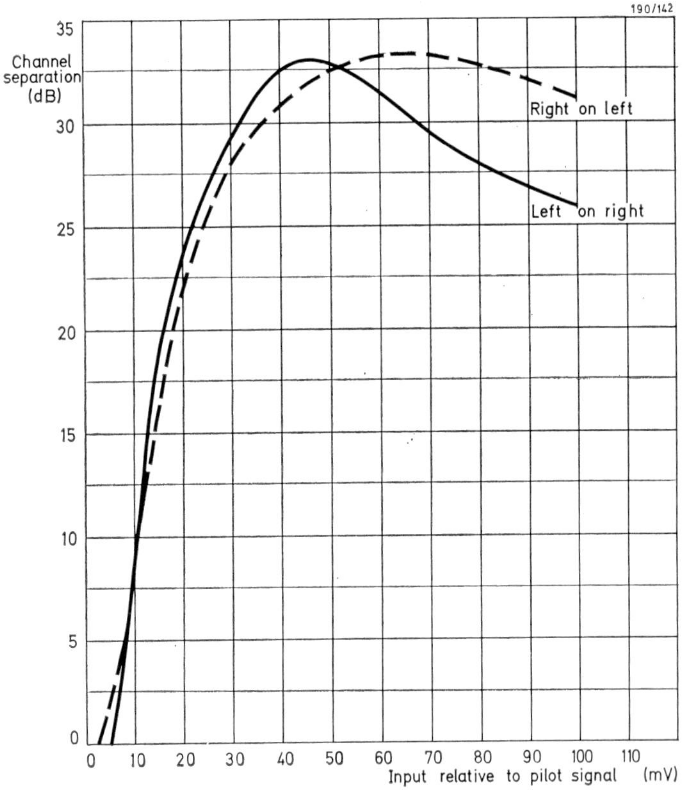

Fig. 142—Channel separation characteristics

signal is of constant amplitude and does not vary with the modulation. Therefore, it is a measure of the signal strength fed to the f.m. tuner. Because its amplitude is much less than that of the other signals, the 19kHz signal is also nearer the noise level. When it is weak the separation between the two a.f. signals is unlikely to be great enough or constant, and the stereophonic effect will be lost.

The output circuit or i.f. filtering circuit of the ratio detector in an f.m. mono receiver requires an a.f. bandwidth of about 15kHz. In an f.m. stereo receiver, however, an a.f. bandwidth of 53kHz is required. Consequently, if all other factors are equal, the signal-to-noise ratio of a stereo receiver is not so good as that of a mono receiver. Therefore, before fitting a stereo decoder, it would be wise to check that the strength of the received signal is sufficient to ensure that the deterioration in signal-to-noise ratio does not make good stereo reception an impossibility.

This check may be made by means of the simple circuits shown in Fig. 143. The circuit in Fig. 143a will produce an attenuation of about 12dB. If this is connected between the aerial feeder and the tuner aerial sockets when receiving mono, there should be little audible difference. Any increase in the background noise (hiss, clicks and pops) indicates that a more efficient aerial or a re-siting of the aerial is needed.

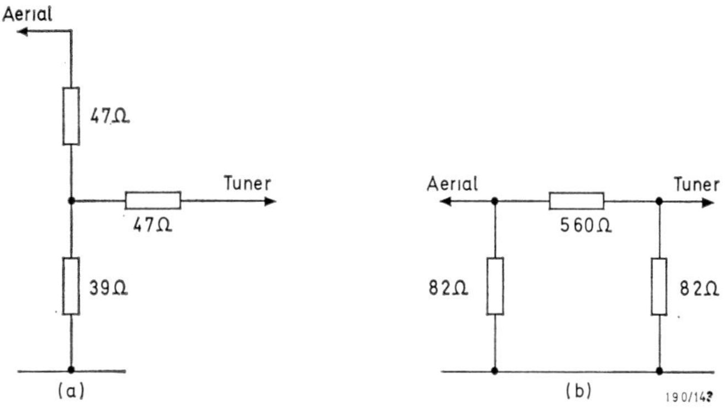

Fig. 143—Circuits for checking signal-to-noise ratio

If the test is satisfactory, two of the circuit in Fig. 143a in series, or one of the circuit in Fig. 143b, should be connected between the aerial feeder and the tuner aerial sockets. If again there is no marked increase in the background noise, the aerial is quite adequate for f.m. stereo. However, if there is a marked increase in the background noise, the aerial system may not be good enough for f.m. stereo. An aerial array or a new position for the aerial may be needed.

COMPONENT LISTS FOR CIRCUITS IN CHAPTER 8

F.M. Tuner using BF200 Transistor (page 202)

Transistors

Circuit reference	Mullard type
TR_1	BF200
TR_2	BF195
TR_3	BF195

Resistors

Mullard CR25 series (0·3W) ±5%

Circuit reference	Value	Circuit reference	Value
R_1	1·5kΩ	R_6	3·3kΩ
R_2	10kΩ	R_7	100Ω
R_3	6·8kΩ	R_8	2·7kΩ
R_4	8·2kΩ	R_9	6·8kΩ
R_5	6·8kΩ	R_{10}	10kΩ

Capacitors

Circuit reference	Value	Description	Mullard type
C_1	3·3pF	tubular ceramic	
C_2	3 to 14pF	ganged tuning capacitor	
C_8	3 to 14pF		
C_{20}	3 to 14pF		
C_3	1 to 7pF	trimmer	
C_4	8·2pF	ceramic plate	C333CB/L8E2
C_5	470pF	tubular ceramic	
C_6	3·9pF	ceramic plate	C333CB/L8E9
C_7	1nF	polystyrene	C295AC/1K
C_9	1 to 7pF	trimmer	
C_{10}	8·2pF	ceramic plate	C333CB/L8E2
C_{11}	1nF	polystyrene	C295AC/1K
C_{12}	22nF	metallised foil	C280AE/P22K
C_{13}	120pF	ceramic plate	C333CH/C120E
C_{14}	1·5nF	polystyrene	C295AA/1K5
C_{15}	82pF	ceramic plate	C333CH/C82E
C_{16}	1pF	tubular ceramic	
C_{17}	1nF	polystyrene	C295AC/1K
C_{18}	1nF	polystyrene	C295AC/1K
C_{19}	1·8pF	tubular ceramic	
C_{21}	1 to 7pF	trimmer	
C_{22}	8·2pF	ceramic plate	C333CB/L8E2

Coils

All the coils are wound with enamelled-copper wire and all except L_5 are air-cored. The i.f. coil, L_5, follows normal design practice to match the input of the amplifier.

Circuit reference	L_1	L_2	L_3	L_4	L_5
Q_O	200		200	200	150
		150			
Q_L	50		100	—	—
No. of turns	4	22	4	4½	—
Wire diam. (mm)	1	0·2	1	1	—
Pitch (mm)	2	close	2	2	—
Inside diam. (mm)	8	4	8	8	—
No. of turns between tap and chassis end of coil	—	—	¾	3½	—

F.M. Tuner using BFW10 Field-effect Transistor (page 212)

Transistors

Circuit reference	Mullard type
TR_1	BFW10
TR_2	BF195
TR_3	BF195

Resistors

Mullard CR25 series (0·3W) ±5%

Circuit reference	Value
R_1	22kΩ
R_2	27kΩ
R_3	680Ω
R_4	8·2kΩ
R_5	6·8kΩ
R_6	4·7kΩ
R_7	100Ω
R_8	3·9kΩ
R_9	6·8kΩ
R_{10}	10kΩ

Capacitors

Circuit reference	Value	Description	Mullard type
C_1	3·3pF	tubular ceramic	
C_2	1·5nF	tubular ceramic	
C_3	1·5nF	tubular ceramic	
C_4	3 to 14pF }	ganged tuning	
C_{15}	3 to 14pF }	capacitor	
C_5	1 to 7pF	trimmer	
C_6	8·2pF	ceramic plate	C333CB/L8E2
C_7	1nF	polystyrene	C295AC/1K
C_8	22nF	metallised foil	C280AE/P22K
C_9	120pF	ceramic plate	C333CH/C120E
C_{10}	1·5nF	polystyrene	C295AA/1K5
C_{11}	1·5pF	tubular ceramic	
C_{12}	22pF	tubular ceramic	
C_{13}	1nF	polystyrene	C295AC/1K
C_{14}	1nF	polystyrene	C295AC/1K
C_{16}	1 to 7pF	trimmer	
C_{17}	8·2pF	ceramic plate	C333CB/L8E2

Coils

All the coils are wound with enamelled-copper wire. L_1 and L_2 are wound on powdered iron; L_3 and L_4 are air-cored. The i.f. coil, L_5, follows normal design practice to match the input of the amplifier.

Circuit reference	L_1	L_2	L_3	L_4	L_5
Q_O	80	80	200		150
Q_L	4	4	100		
No. of turns	3	4	4	4½	
Wire diam. (mm)	0·15	0·15	1	1	
Pitch (mm)			2	2	
Inside diam. (mm)			8	8	
No. of turns between tap and chassis end of coil			drain 1 mixer ½	3½	

Stereo Decoder and Indicator (pages 220 and 222)

Transistors and Diode

Circuit reference	Mullard type
TR_1	BC148 or BC108
TR_2	BC148 or BC108
TR_3	BC148 or BC108
TR_4	BC148 or BC108
TR_5	BC147 or BC107
D_1	OA91

Resistors

Mullard CR25 series (0·3W) ±5%

Circuit reference	Value	Circuit reference	Value
R_1	22kΩ	R_{14}	56kΩ
R_2	10kΩ	R_{15}	12kΩ
R_3	10kΩ	R_{16}	12kΩ
R_4	10kΩ	R_{17}	1·8kΩ
R_5	220Ω	R_{18}	1·8kΩ
R_6	4·7kΩ	R_{19}	1·8kΩ
R_7	100Ω	R_{20}	1·8kΩ
R_8	1kΩ	R_{21}	18kΩ
R_9	3·9kΩ	R_{22}	18kΩ
R_{10}	1kΩ	R_{23}	150kΩ
R_{11}	1kΩ	R_{24}	150kΩ
R_{12} preset potentiometer	1kΩ	R_{25}	5·6kΩ
R_{13}	100kΩ	R_{26}	47kΩ

Capacitors

Circuit reference	Value	Description	Mullard type
C_1	10nF	metallised foil	C280AE/P10K
C_2	470pF	polystyrene or silvered-mica	
C_3	100nF	metallised foil	C280AE/P100K
C_4	10nF	metallised foil	C280AE/P10K
C_5	470pF	polystyrene or silvered-mica	
C_6	10μF	electrolytic 16V	C426AR/E10
C_7	4·7nF	polyester	C296AC/A4K7
C_8	4·7nF	polyester	C296AC/A4K7
C_9	4·7nF	polyester	C296AC/A4K7
C_{10}	4·7nF	polyester	C296AC/A4K7
C_{11}	10μF	electrolytic 16V	C426AR/E10
C_{12}	2·2nF	polystyrene	C295AA/2K2
C_{13}	2·2nF	polystyrene	C295AA/2K2
C_{14}	2·2nF	polystyrene	C295AA/2K2
C_{15}	2·2nF	polystyrene	C295AA/2K2
C_{16}	220pF	ceramic plate	C333CH/C220E
C_{17}	220pF	ceramic plate	C333CH/C220E
C_{18}	100nF	metallised foil	C280AE/P100K
C_{19}	100nF	metallised foil	C280AE/P100K
C_{20}	10μF	electrolytic 25V	C426AR/H10
C_{21}	10nF	metallised foil	C280AE/P10K

Capacitors C_2 and C_3 should be high-quality polystyrene or silvered-mica types.
Capacitors C_{12} to C_{17} should be ±5% components.

Coils

Circuit reference	Mullard type
L_1	WF2949
L_2/L_3	WF2951

For coil details see Figs. 144 and 145.

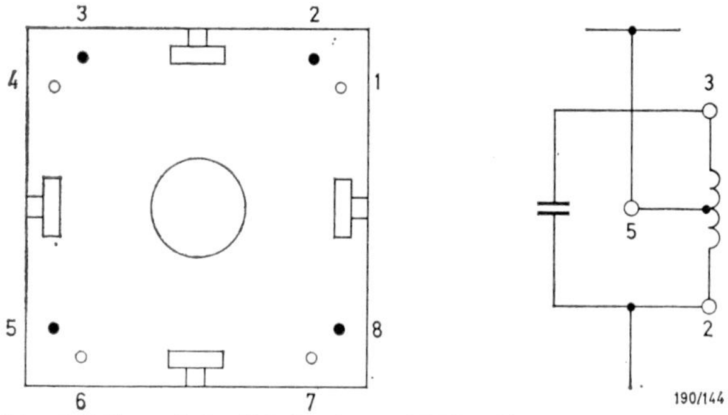

Fig. 144—19kHz coil, L_1 (Mullard type WF2949), viewed from underside. Pin 8 is the coil clamp and should be connected to earth

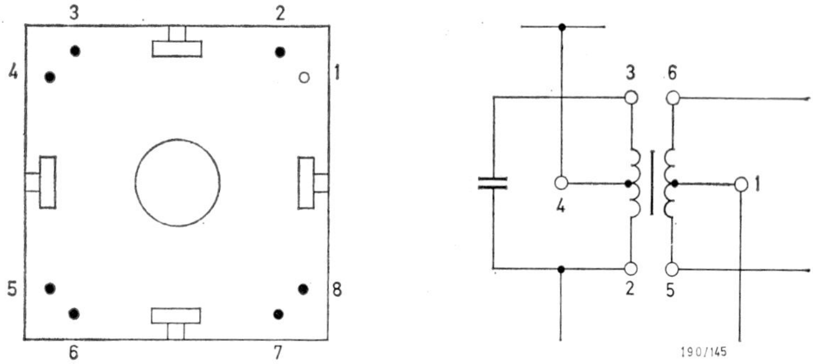

Fig. 145—38kHz coil, L_2/L_3 (Mullard type WF2951), viewed from underside. Pin 8 is the coil clamp and should be connected to earth

Stereo Indicator Bulb LPI

A 24V, 20mA indicator bulb is required, a suitable component being the Electroniques Limited type SGF9/D/GN/24, available from Edinburgh Way, Harlow, Essex.

CHAPTER 9

LOUDSPEAKER SYSTEMS

The choice of loudspeaker for a particular sound-reproduction system depends on many factors. Some of these factors are discussed in this chapter, together with the methods of mounting loudspeakers and the design of enclosures. Design details of enclosures are not given here; such details can be found in specialist journals and books.

THE LOUDSPEAKER

The diagrammatic cross-section of a typical moving-coil loudspeaker is shown in Fig. 146. The speech coil is suspended in the gap of a powerful permanent magnet by means of the centring device or 'spider', and is attached to the small end of the cone. The wide end of the cone is attached to the frame by the flexible cone surround.

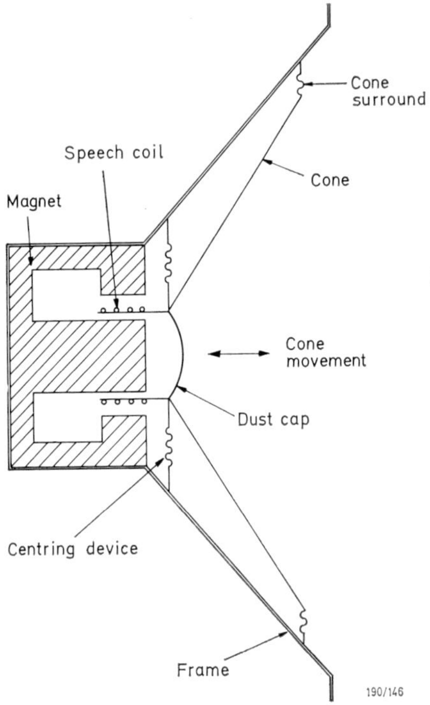

Fig. 146—Diagrammatic cross-section of typical moving-coil loudspeaker

The audio-frequency output signal from the amplifier is applied to the speech coil. The current flowing in the coil produces a magnetic field which reacts with the field of the permanent magnet and so moves the loudspeaker cone. Thus the sound waves produced by the movement of the cone are a 'reproduction' of the audio-frequency signal from the amplifier.

Great care has to be taken over the mechanical design of the loudspeaker to prevent unnecessary distortion being introduced into the sounds produced. The advantages of providing an electrical signal with the minimum distortion throughout the various stages of amplification will be lost if an inferior loudspeaker is used.

The main causes of distortion are inherent in the principle of the loudspeaker itself—the fact that a movable cone is used to launch sound waves in response to an electrical signal. The relationship between the force on the cone (produced by the speech coil) and the displacement of the cone is not linear. This gives rise to distortion which increases as the cone displacement increases. Overloading the speaker will therefore produce distortion. At high frequencies where the wavelengths of the sounds produced are smaller than the depth of the cone, interaction of the launched sound waves may produce frequencies that were not present in the original electrical signal applied to the speaker. Again, a loudspeaker produces sounds of several frequencies at the same time and these different frequencies can interact to produce unwanted frequencies. This effect is known as intermodulation distortion. Another form of distortion is Doppler distortion, caused by cone movements relating to one frequency themselves being moved by cone movements in response to other frequencies. Sound energy can be reflected back along the cone from the edge. The reflected waves will interact with the original sound waves, again producing unwanted frequencies.

Some of these types of distortion can be eliminated or reduced by good mechanical design and the use of special materials. For example, distortion caused by reflections from the edge of the cone can be minimized by absorbing the sound at the edge by using a suitable material for the cone surround. Unfortunately, the requirements for good reproduction of one group of frequencies may be in conflict with those for another group of frequencies. Thus (as will be discussed later), the design of a wide-range speaker must be a compromise, or more than one speaker must be used to cover the required frequency range.

It is worth remembering that 'sound quality' is subjective. The ear is more tolerant of distortion at low frequencies than at high frequencies. A higher distortion can be tolerated on a stereo system than on a mono system. And personal preferences cannot be ignored. What is considered a good treble response to one listener may appear shrill to another, and one man's 'bags of bass' is another man's 'boom'.

FACTORS AFFECTING CHOICE OF LOUDSPEAKER

Because the loudspeaker converts electrical energy into acoustic energy, the ratio of the acoustic energy produced to the electrical energy supplied gives the efficiency of the speaker. For high-quality systems, low distortion in the speaker is more important than high efficiency, and so efficiencies of 1 or 2% are acceptable. Higher efficiencies are necessary for the speakers used with, say, portable radios and record players to avoid a high power output from the amplifier with the consequent problems of heat dissipation in the output transistors and high battery drain. Thus one factor affecting the choice of loudspeaker is the efficiency required for the particular application.

The range of frequencies that can be reproduced by the speaker is another factor to be considered. The relative acoustic power levels occurring in speech and music at different frequencies are shown in Fig. 147. It can be seen that there is a higher power level in the low frequencies than in the high frequencies, and so for good sound reproduction a higher power is required for the bass frequencies. For realistic bass reproduction, a large-diameter speaker is required, with a cone of high mass. On the other hand, for good high-frequency reproduction a small-diameter speaker is required (to prevent interaction of the launched sound waves), with a cone of low mass. The cone of the high-frequency speaker can also be shaped to reduce this interaction, a 'dome' rather than a 'cone'. Thus the requirements for good high-frequency and good low-frequency reproduction are in conflict.

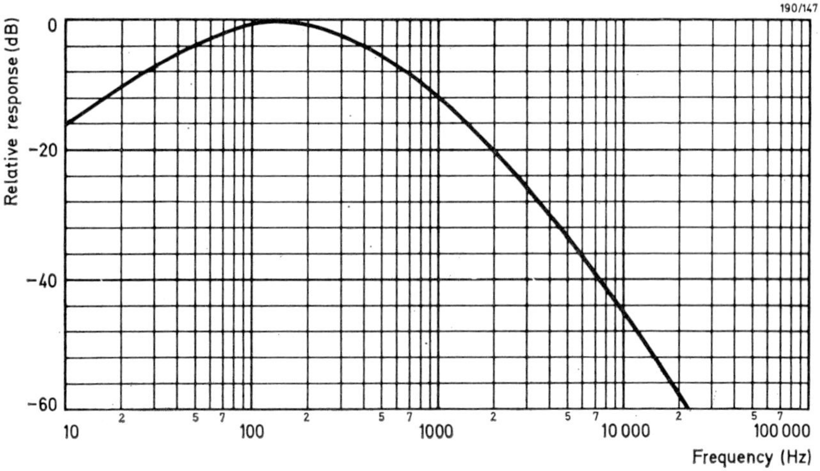

Fig. 147—Relative acoustic power in speech and music plotted against frequency

If the frequency range of the speaker is not too wide, a single speaker can be used whose mechanical design is a compromise between the conflicting requirements. Another solution is to use a speaker with a good bass response but take advantage of the fact that at high frequencies the speech coil may be the only part of the speaker to be moving, and that the acoustic energy is radiated mainly from the apex of the cone. An additional small cone, lightweight but stiff, can be attached to the apex of the main cone to radiate the high frequencies more efficiently. The speech coil itself can be made very light by winding it from aluminium wire.

For high-quality systems, the solution to this problem of conflicting requirements is to use two speakers, each specially designed to handle a range of frequencies with minimum distortion. A large-diameter speaker is used for the bass frequencies, while a small-diameter speaker with a lightweight cone is used for the treble frequencies. A crossover network is used to divide the complete frequency range between the two speakers.

A typical crossover network is shown in Fig. 148. It consists essentially of a high-pass filter in series with the high-frequency speaker, and a low-pass filter in series with the low-frequency speaker. The values of the

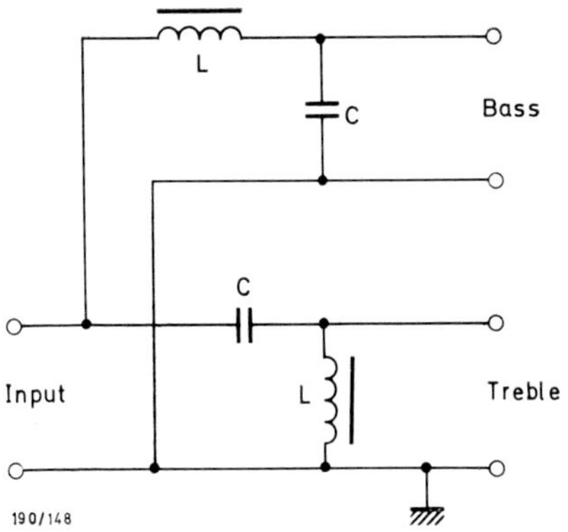

Fig. 148—Crossover network for two speakers

inductor L and the capacitor C are given in terms of the loudspeaker impedance R and crossover frequency f by:

$$L = \frac{\sqrt{2}.R}{2\pi f}, \text{ and } C = \frac{\sqrt{2}}{4\pi fR}.$$

The overall sound quality from a speaker system using two speakers is better than that of a single-speaker system although, of course, the two-speaker system is more expensive. Systems using three or even more speakers are possible. The speaker for the bass frequencies has acquired the name 'woofer' while that for the treble frequencies is called the 'tweeter'. If a third speaker is used to cover the middle frequencies, this is known as the 'squawker'.

The frequency response of a speaker is measured by supplying the speaker with a constant-voltage signal of varying frequency and recording the output under standard conditions. The measurement is carried out in an anechoic room, with the speaker mounted 'free' (that is, not mounted on a baffle or in an enclosure). The speaker output is recorded by a microphone placed a specified distance along the speaker axis. The amplitude of the microphone signal as the frequency of the signal supplied to the speaker is swept through the range can be used to plot the response. The frequency range used is usually from 20Hz to 20kHz, and the response is expressed as a relative level, the 0dB level being 52dB above the threshold of hearing (2×10^{-4}μbar or 20μN/m^2).

A typical 'free space' speaker response is shown in Fig. 149, the curve shown being for an $8\frac{1}{2}$in wide-range speaker. The absence of any mounting for the speaker is clearly shown by the apparently poor low-frequency response. The fall in the response of low-frequencies begins as high as 2kHz. Mounting the speaker on a baffle or in an enclosure radically alters this part of the response.

Another problem associated with speakers reproducing a wide frequency range is directional response. Bass frequencies are well diffused over the listening area, but as the frequency is increased, the sound becomes

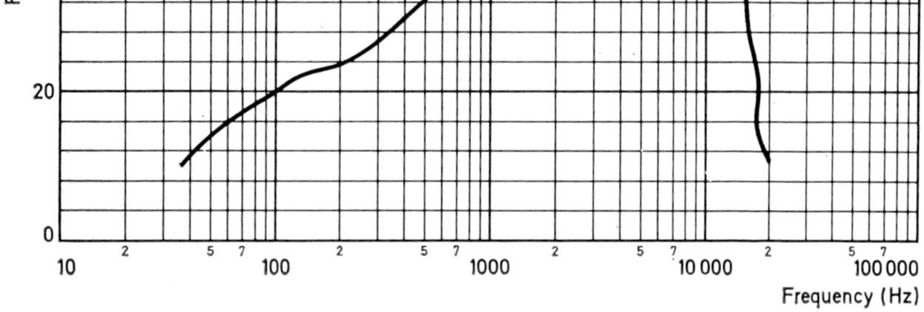

Fig. 149—Typical frequency response for wide-range speaker unmounted

noticeably directional. Thus the strength of the treble sounds will decrease away from the speaker axis, as shown diagrammatically in Fig. 150. The directional effect of the treble frequencies detracts considerably from the realism of the reproduced sounds for a listener not sitting along the axis of the speaker.

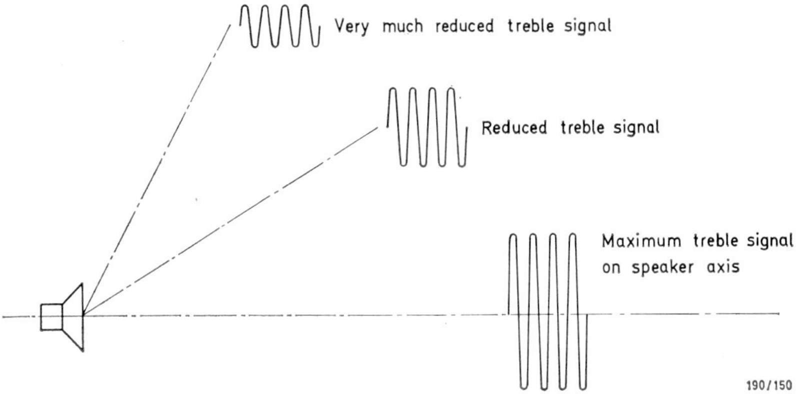

Fig. 150—Directional effect of treble frequencies from speaker

Again this problem can be overcome by using a multi-speaker system. With mono systems, the tweeter can be mounted so that the treble frequencies are diffused before reaching the listening area, for example by bouncing the sound off the ceiling of the listening room. With stereo systems where directional information is required to produce the stereo illusion, the spread of the sound waves obtained from modern tweeters is often sufficient to give an adequate listening area.

If a speaker is placed in an anechoic room as for the frequency response measurement but with a constant-frequency constant-voltage signal applied, and the speaker is rotated while the microphone remains stationary, the polar response curve is obtained. A typical polar response curve is shown in Fig. 151. The 0° corresponds to the speaker axis, and the figures round the outside of the graph correspond to the angle away from the axis. The relative response of the speaker for the particular frequency is shown by the distance of the curve from the centre of the graph.

The curve of Fig. 151 is for the same wide-range speaker used for the frequency response curve in Fig. 149. The polar response is reasonably uniform at the frequency used, 2kHz, but a second polar response for

Fig. 151—Polar response of wide-range speaker at 2kHz

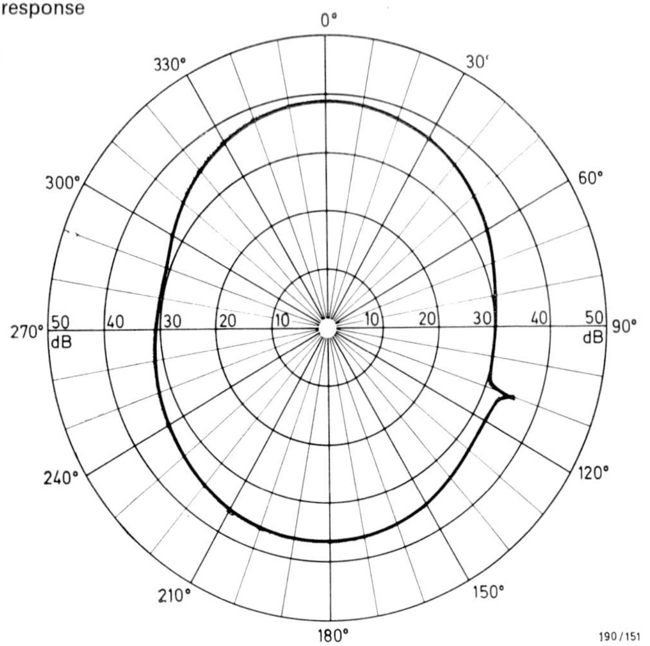

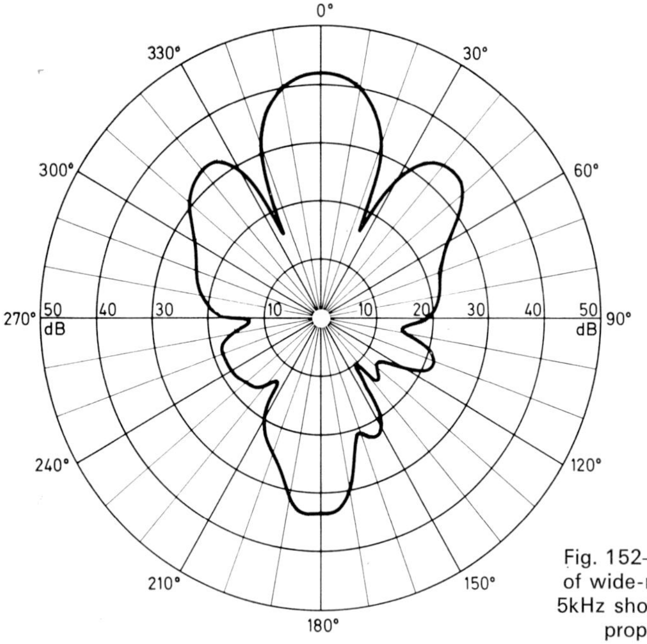

Fig. 152—Polar response of wide-range speaker at 5kHz showing directional properties of speaker

5kHz, Fig. 152, shows the directional properties of the speaker appearing.

Every loudspeaker has a fundamental resonance frequency at which a very high acoustic output is produced. Good sound reproduction cannot normally be obtained below the resonance frequency because every time the electrical signal drives the speaker through the resonance frequency, an unnaturally loud sound would be produced. The resonance frequency of the speaker must therefore be as low as possible. However, certain types of enclosure damp this resonance so that the speaker can be used satisfactorily below the resonant frequency.

The speech coil (in its simplest form) possesses both resistance and inductance, and the inductive reactance will vary with the frequency of the signal from the amplifier. Thus the speaker impedance and the current drawn from the amplifier will vary with frequency. The nominal impedance quoted for a speaker is the lowest measured value above the resonance frequency. Thus for an 8Ω speaker, the value of 8Ω is the lowest measured above the resonance frequency, but the impedance at other frequencies can be two or three times this value. In a multi-speaker system, the impedance of the components forming the crossover network will also be presented to the amplifier, although this usually results in a peak in the impedance characteristic. The variation in impedance with frequency for a high-quality speaker system using two speakers with a crossover frequency of 3·5kHz is shown in Fig. 153.

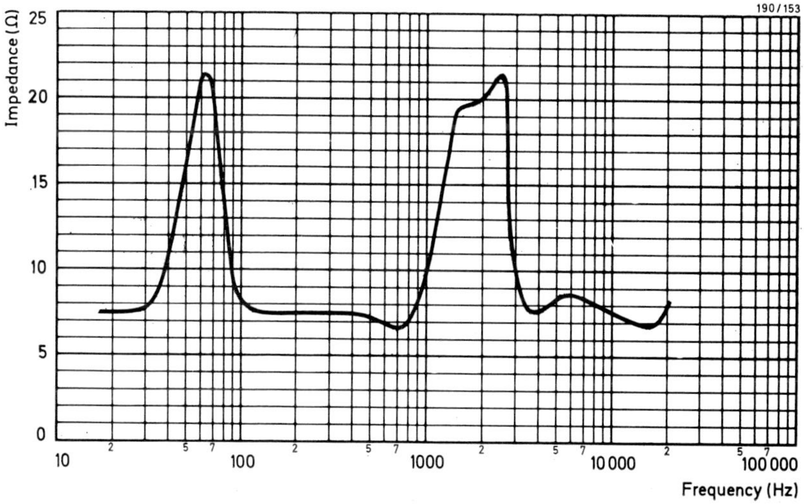

Fig. 153—Variation of impedance with frequency for two-speaker system with crossover frequency of 3·5kHz

The nominal impedance quoted for the speaker is the lowest value because this is the value which must be matched to the rated load impedance for the amplifier. If the difference between the speaker impedance and the amplifier rated load impedance is too great there will be a loss in power, but if the speaker impedance is much lower than that of the amplifier then the amplifier may be damaged.

The power-handling capability of the speaker in relation to the amplifier must also be considered. A 10W speaker can be connected safely to the output of a 10W amplifier, although it is permissible to connect the speaker to an amplifier with a lower output power as this allows a safety margin. What must not be done is to connect a speaker to an amplifier of higher power since the speaker may be permanently damaged if full output power is reached. It should be remembered that the loudspeaker power rating, as with the output power of an amplifier, does not imply that the speaker always requires that power for normal drive. It is a measure of the maximum power that can be handled safely, for example during a loud passage of music or during a short rapidly-rising loud transient.

The factors affecting the choice of loudspeaker that have now been discussed in some detail may be summarized into two groups. First there are the factors that are common to all loudspeakers and which the designer cannot alter. These are the impedance and power rating which must be matched to the rated load impedance and power rating of the amplifier driving the speaker. The second group of factors are those which allow the designer a choice to meet the needs of the particular application. These are: frequency range, depending on the signal source and amplifier; resonance frequency, which must be below the lowest frequency to be reproduced; efficiency, depending on the output power of the amplifier; and the quality/economic considerations for the application such as the sound quality required, whether one or two speakers can be used, quality of the signal sources that are likely to be used, and so on.

CLASSIFICATION OF LOUDSPEAKERS AND APPLICATIONS

A convenient classification of sound-reproducing systems from the point of view of the loudspeaker used is into three groups:

'economical' systems,
standard systems,
high-quality systems.

The type of loudspeaker for each of these groups will be discussed below.

Economical Systems

An 'economical' system is one in which the selling price of the equipment is considered to be of more importance than the quality of the reproduced sound. Equipment falling into this category includes small portable a.m. transistor radios, portable record players, and small tape recorders. The equipment is compact, and so the space available for the speaker will be limited. Thus a small speaker will have to be used, with a consequent limit on the sound quality. Also, because the cabinet in which the speaker is mounted is small, there will be a limited low-frequency response and so the speaker itself need not have an extended low-frequency range.

Because the output power from the amplifier will be low, a speaker with a high efficiency is required. The frequency range required of the speaker will be restricted, and so the resonance frequency can be comparatively high.

The loudspeakers for this type of equipment would be circular speakers with diameters from $1\frac{1}{2}$ or 2in to 5in, or elliptical speakers up to approximately 6 by 4in. The frequency response required would be from 200Hz to 10kHz (say), and the speaker resonance frequency would be between 100 and 300Hz.

Standard Systems

The standard systems include all domestic sound-reproducing equipment with the exception of high-quality systems. Mains a.m./f.m. radios, radiograms, and record players with separate speakers come into this group. The cabinets of the equipment are more substantial than those of the first group, and so the cabinet will act to some extent as a baffle. This has the effect of extending the frequency response of the speaker (as is discussed later), and a similar improvement is obtained when the speakers of a record player are mounted in separate enclosures. Generally a single wide-range speaker is used with these systems, although sometimes a tweeter may be added.

The speakers used will have diameters from 5 to 10in. The frequency response required will be from 60Hz to 15kHz (say), and the speaker resonance frequency between 50 and 100Hz.

High-quality Systems

The speakers used in high-quality audio systems are always mounted in specially designed enclosures. The enclosure ensures that the speaker is always properly loaded, by the column of air that is displaced by the movement of the cone in response to the signal from the amplifier, and so can safely handle its rated power. The correct loading is provided by the volume of the cabinet, and unless an enclosure of the correct size is used, there is danger of damaging the speaker on peak powers. The enclosure can be airtight or tuned to a particular frequency by a port. Airtight enclosures have the advantage of being smaller than tuned enclosures.

Multi-speaker enclosures are used for high-quality systems, generally with either two or three speakers. In high-power systems, or very-high-quality systems such as studio monitors, more than one tweeter or squawker may be used.

The speakers used as woofers have dimensions between 5 and 12in, with resonance frequencies between 25 and 50Hz to give a frequency range down to 30Hz or below. The tweeters are 1 or 2in in diameter with specially profiled cones, or may be in the form of small horns. Typical resonance frequencies are between 800 and 1000Hz, and the high-frequency response extends to 20kHz. The speakers used as squawkers are typically 5in diameter with a resonance frequency of 250Hz.

MOUNTING OF LOUDSPEAKERS

In the previous sections mention has been made of the mounting of speakers. The method of mounting has a considerable effect on the behaviour of the speaker, and this will be considered in this section. Strictly speaking, the discussion applies only to high-quality speakers, although some of the conclusions will apply to the standard systems.

Simple Baffle Mounting

The polar response curve of an unmounted loudspeaker at 500Hz is shown in Fig. 154. It can be seen the response falls considerably at 90° and 270° to the speaker axis, caused by 'acoustic short-circuiting'. This occurs when the depth of the speaker corresponds to one-half the wavelength of the sound waves produced, and is caused by interaction between the sound waves produced by the front of the cone and those produced by the back.

This interaction can be considerably reduced by mounting the speaker on a large panel so that the sound waves produced by the two sides of the cone are separated. Such a panel is called a baffle. Mounting the speaker on a baffle also improves the low-frequency response. The overall frequency response of an unmounted speaker (such as that shown in Fig. 149) can be simplified to the curve shown by the full line in Fig. 155. The middle-frequency response is level, but the response starts to fall at frequency f_1 as the half-wavelength of the sound waves produced becomes comparable with the depth of the speaker. The response falls even more rapidly below the speaker resonance frequency f_0. Mounting the speaker on a baffle modifies the response to that shown by the broken line. The level part of the response is extended to frequency f_2, and the response at the resonance

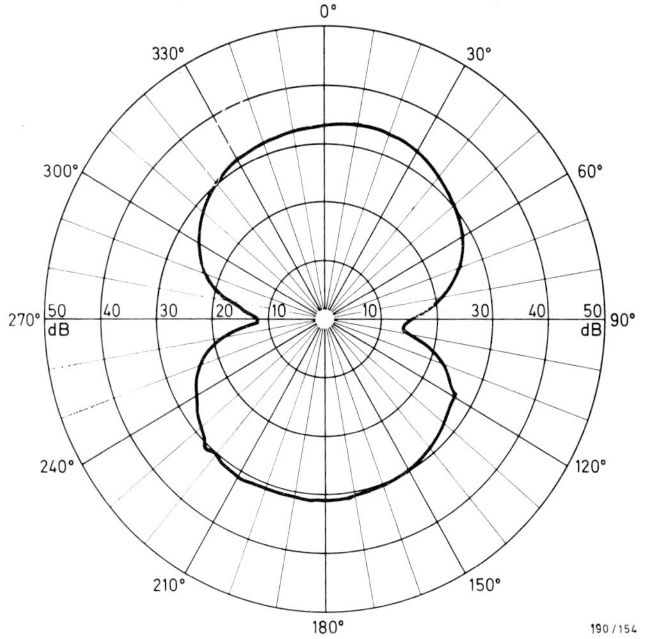

Fig. 154—Polar response of speaker at
500Hz showing effect of 'acoustic short-circuiting'

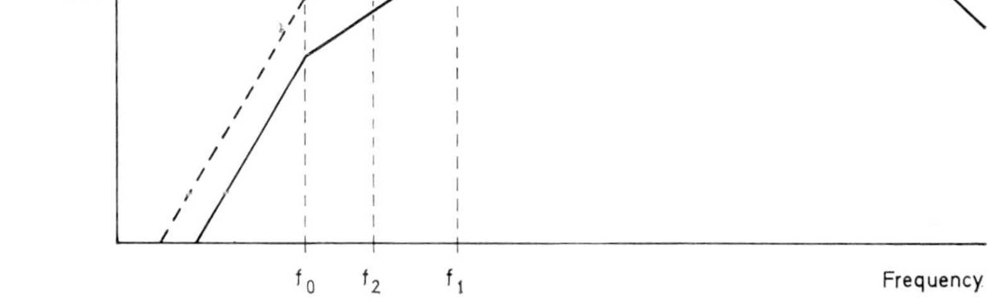

Fig. 155—Simplified response curve of speaker showing effect of baffle mounting

frequency is higher. Thus the bass response of the speaker is improved.

The baffle does not completely prevent interaction between the sound waves from the front and back of the cone, but the interaction starts at a frequency depending on the dimensions of the baffle. The lowest frequency that can be produced by the baffle without interaction has a half-wavelength equal to the shortest dimension of the baffle. Thus to produce a frequency of 40Hz, wavelength 8·5m or 28ft approximately, a baffle of 4·25m (14ft) would be necessary. In practice, the size of the baffle can be reduced because of the area of the floor on which it stands, and the area of the walls to which it is attached. However, baffle mounting of speakers is generally not practicable for domestic systems because of the large size necessary for the baffle if good bass response is to be obtained.

Reflex Enclosures

To reduce the space required for baffle mounting, the baffle itself can be folded into a box. If a port is cut in the box, the enclosure will act as a resonating system because of the behaviour of the air at the port. The resonant frequency depends on the volume of the enclosure and the dimensions of the port. This type of enclosure is called a reflex enclosure, and the resonant frequency of the enclosure is matched to the speaker resonance frequency to ensure correct loading and a good bass response.

Reflex enclosures were extensively used in the early days of Hi-Fi, and are still used today. However, the smaller size possible with the sealed enclosure has led to the replacement of reflex enclosures for domestic use.

Sealed Enclosures

If the baffle is again folded into a box, but no port cut in the enclosure, then an 'infinite baffle' is obtained. If the enclosure is made airtight, separation of the sound waves produced by the front and the back of the speaker cone is achieved. This type of enclosure has the advantages over the reflex type of generally being smaller and not requiring special matching to the speaker by a tuned port. As long as the volume of air within the enclosure is correct, the speaker will be properly loaded. The relative dimensions of the enclosure can be varied so long as the correct volume is maintained, and this allows wall-mounted and bookshelf enclosures to be manufactured as well as the familiar floor-standing type. The disadvantage of the sealed enclosure is that the bass response is slightly inferior to that of the reflex enclosure.

Sealed enclosures will generally contain more than one speaker. The loading on each speaker must be correct, and so the tweeter and squawker (if used) must be contained in separate sealed enclosures. Thus a high-quality speaker enclosure will consist of one or more sealed enclosures contained in the sealed woofer enclosure. Allowance must be made for these other enclosures when designing the overall enclosure to ensure that the remaining volume of air provides the correct loading for the woofer. The only limitation on the dimensions is that the depth of the enclosure should be greater than $1\frac{1}{2}$ times the depth of the speaker. For a multi-speaker enclosure, the depth will be determined by the depth of the woofer.

It is important that the enclosure should be as rigid as possible, and internal struts are often used for additional stiffness. It is important that the enclosure does not vibrate, otherwise the sound produced may be 'coloured'. To prevent internal reflection of the sound waves and the formation of standing waves, the inside of the enclosure is lined with some sound-absorbent material.

The mounting of the speakers in the enclosure is not critical. Generally the woofer is mounted near the bottom of the enclosure, mainly because it is the heaviest speaker and gives the enclosure stability. Nor does it matter whether the enclosure is horizontal or vertical, thus allowing the same enclosure to be used as a floor-standing or bookshelf unit. The only limitation on the layout of speakers in an enclosure occurs when two speakers are used for the same range of frequencies. Thus if two tweeters are used, they should not be placed too close together horizontally. Vertical stacking of woofer, squawker, and tweeter is generally used as any interaction of the sound produced will occur in the vertical plane, and will not be noticed by the listener unless he moves his head vertically.

The frequency response of a typical high-quality sealed enclosure is shown in Fig. 156. It can be seen that the response is flat from approximately 100Hz to 20kHz. This response should be compared with those for the unmounted speakers used in the enclosure shown in Fig. 157 from which it can clearly be seen how the enclosure affects the low-frequency responses.

ROOM ACOUSTICS

No discussion of loudspeakers and enclosures would be complete without some consideration of the listening room acoustics. However good the sound produced by the speakers, the quality of the sound reaching the listener's ears will depend on the acoustic properties of the room.

The sound heard by the listener is a mixture of direct sound from the loudspeakers and sound reflected from the walls, ceiling, and furniture in the room. In an anechoic room, only direct sound is heard as the sound reaching the walls, ceiling, and floor is absorbed. This absorption produces

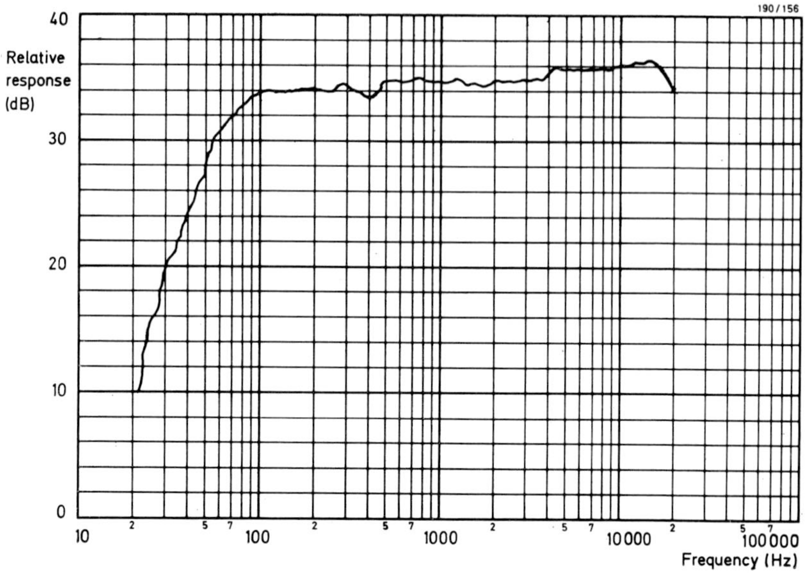

Fig. 156—Frequency response of high-quality speaker system in sealed enclosure

the characteristic dead sound of an anechoic room. At the other extreme is the large hall with highly reflective walls and ceiling. The sound is reflected from these surfaces and takes a long time to decay. If the hall is large enough there may be an echo. Even if this stage is not reached in a domestic room, excessive reflection may affect listening, particularly the clarity of quick passages of music.

A measure of the acoustic properties of a room or hall is the reverberation time. This is defined as the time taken for a sound to decay to one-millionth of its original energy level. The reverberation time for an average domestic room is about half a second. The sound is absorbed by the carpet, curtains, and furnishing fabrics, and so some control over the acoustic properties of the room can be exercised by the amount and positioning of these absorbing materials.

A problem that may be experienced is the formation of standing waves. When one of the dimensions of the room corresponds to half the wavelength of the particular sound being reproduced by the speaker, resonance will occur. Because this can happen in the vertical plane as well as in the

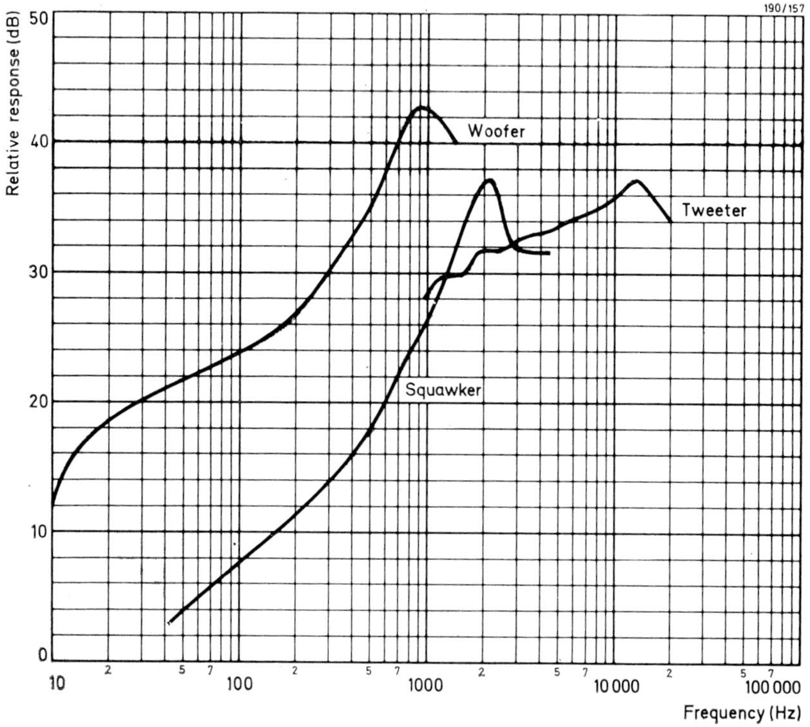

Fig. 157—Combined unmounted frequency responses for speakers used in the system of Fig. 156

horizontal plane, and can occur with harmonics as well as the fundamental, the situation is complex. If a standing wave pattern occurs at certain frequencies, it is shown by changes in sound level and quality in different parts of the room. The problem can generally he overcome by altering the relative positions of the speakers and furniture in the room, particularly to avoid large reflective surfaces.

The best position for the speakers can only be found by experience. Generally a corner position is used, but this is not always the best position. With a stereo system, the speakers should not be placed too far apart otherwise sound with 'a hole in the middle' may result. Unless the room can be arranged around the speakers, some compromise arrangement will have to be found by trial-and-error. Bookcase and wall-mounted speakers should not be mounted too high—the best height for a speaker is about ear level when the listener is sitting down.

Finally, it should be remembered that the lowest frequency that can be satisfactorily reproduced depends on the room diagonal from floor corner to the opposite ceiling corner. This dimension represents half the wavelength of the lowest frequency that can be sustained by the room. Frequencies below this cut-off value cannot be reproduced.

Table 20 gives a list of frequencies and the corresponding room diagonal equivalent to the half-wavelength. The room diagonal itself can be calculated from Pythagorus's Theorem. The room diagonal D is given by:

$$D^2 = L^2 + W^2 + H^2,$$

where L is the length of the room, W the width, and H the height.

TABLE 20

Cut-off frequencies for corresponding room diagonal

Cut-off frequency (Hz)	Room diagonal (ft)	(m)
20	28	8·5
25	22·4	6·8
30	18·5	5·66
35	15·9	4·85
40	14	4·25
45	12·4	3·78
50	11·2	3·4

CHAPTER 10

TEST EQUIPMENT

Items of test equipment which will be found useful in commissioning and maintaining audio and radio circuitry include a voltmeter, ohmmeter, milliammeter, output power meter, a.f. signal generator, i.f. and r.f. signal generator with provision for modulation at audio frequencies, oscilloscope, transistor tester and stereo balance meter.

It is beyond the scope of this book to give designs for the complex equipment required for aligning and testing h.f. circuitry, and it is recommended that professionally built equipment should be obtained for this purpose.

VOLTMETER

It is essential that the voltmeter does not alter the conditions in the circuit when measurements are being made. The voltmeter should therefore have a resistance of at least 20kΩ/V. A voltage coverage of 2·5 to 100V d.c. is required. A valve voltmeter is very suitable for the measurements if precautions are taken to prevent earth-loop currents damaging the transistors.

The simplest approach to the testing of the directly-coupled amplifiers described in this book is to measure voltage levels throughout the circuit. A conventional four-transistor circuit is used here as an example, as shown in Fig. 158, and the voltage levels resulting from various transistor failures are given in Table 21.

TABLE 21

Voltage Levels

Transistor Fault	Voltage at point				General Performance
	A	B	C	D	
TR_2 open-circuit	high	no voltage	high	slightly higher than normal	no output
TR_2 short-circuit	low	twice normal voltage	low	slightly higher than normal	no output
TR_3 open-circuit	near-chassis	low	high	low	no output
TR_4 open-circuit	normal	normal	normal	normal	high distortion
TR_3 or TR_4 short-circuit	indeterminate				emitter resistors burn out

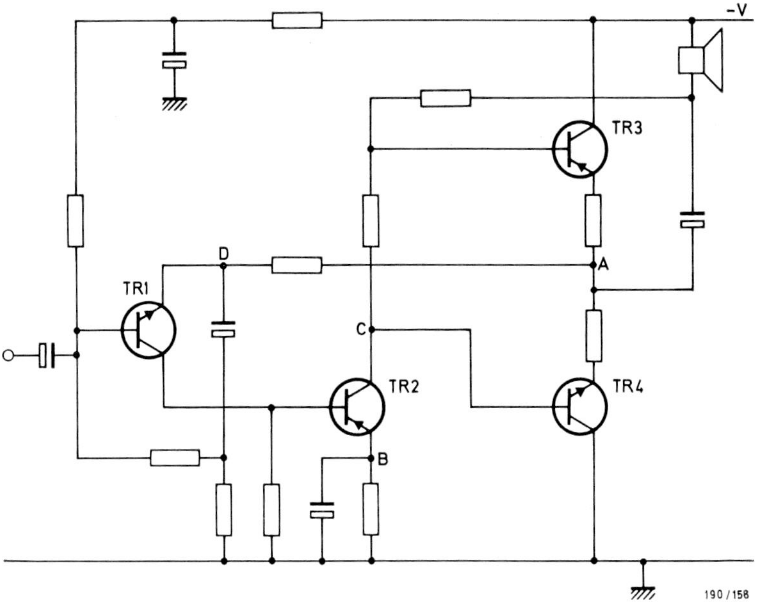

Fig. 158—Direct-coupled amplifier with voltage test points

OHMMETER

The voltage at the output terminals of the ohmmeter should not exceed 1·5V otherwise there is a possibility of the transistors and miniature electrolytic capacitors in the circuit under test being damaged. In all cases when resistance measurements are made, the equipment should be switched off.

If only one component is suspect this can be isolated from the circuit by unsoldering the connection at one end and then testing with the ohmmeter. However, if components have to be checked while they are still in the circuit there are two methods available to prevent the 'diode elements' of the transistors affecting the measurements. In the first method the base connection of the transistor is unsoldered so that the transistor is effectively disconnected from the circuit. The resistance measurements may then be made, although allowance must be made for other resistors shunting the component under test. The second method uses the voltage of the ohmmeter to 'reverse bias' the diode elements of the transistor so that the transistor does not conduct.

The basic circuit of a series-connected ohmmeter is shown in Fig. 159 in which both the colour and the polarity of the output terminals are shown. To provide the reverse bias for a p-n-p transistor, the positive ohmmeter terminal (coloured black) is connected to the base (n-type material) and the negative ohmmeter terminal (coloured red) connected to the emitter or collector (p-type material). The connections for measuring the values of the base bias resistors of a p-n-p transistor are shown in Fig. 160; for an n-p-n transistor the connections are reversed. If the stage being checked is not isolated from the rest of the circuit the other stages will appear as a resistance in parallel with that being measured. The measured value of resistance will therefore be lower than the actual resistance.

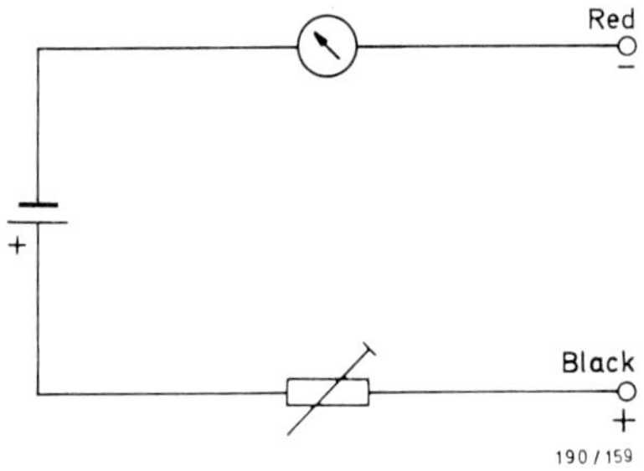

Fig. 159—Circuit of series-connected ohmmeter

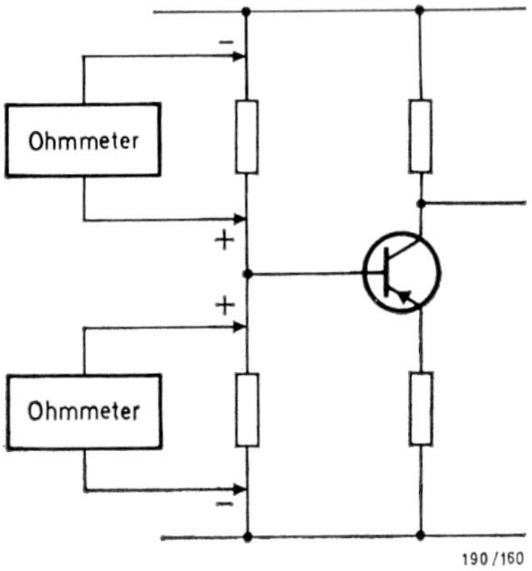

Fig. 160—Method of checking base bias resistors with ohmmeter

MILLIAMMETER

A milliammeter with a full-scale deflection of 100mA may be used to measure the currents in the pre-amplifier and driver stages of the amplifiers described, but the output transistors in some of the circuits draw currents in excess of 1A.

OUTPUT-POWER METER

An output-power meter can be used for checking a.f. output power. The loudspeaker should be replaced by the output-power meter which is switched to the loudspeaker impedance. The meter then indicates the output power directly.

A voltmeter can also be used conveniently for this measurement if an output-power meter is not available. The loudspeaker should be replaced by a resistor corresponding to the loudspeaker impedance and the power level calculated as the voltage drop across the resistor.

AUDIO-FREQUENCY SIGNAL GENERATOR

The signal generator shown in Fig. 161 is a Wien-bridge oscillator which will supply an output voltage of 1V r.m.s. over a frequency range 15Hz to 200kHz.

Circuit Description

The circuit uses a thermistor R_5 as an amplitude control device, and the output is essentially independent of small changes in supply voltage and ambient temperature. The oscillator uses only three transistors and, apart from the frequency-determining capacitors, only one capacitor. The circuit operates from a 9V supply, although little difference in performance is obtained with supply voltages from 6 to 10V. The current consumption is:

6V supply	6·3mA
9V supply	11·5mA
10V supply	13mA

Performance

The frequency coverage is 15Hz to 200kHz in four ranges:

Range 1	15Hz to 200Hz
Range 2	150Hz to 2kHz
Range 3	1·5kHz to 20kHz
Range 4	15kHz to 200kHz

The approximate harmonic distortion at 1kHz is:

second harmonic	0·1%
third harmonic	0·05%

At frequencies above 100kHz the frequency calibration does not coincide with that for lower ranges, because of stray capacitances, and the output voltage is slightly higher.

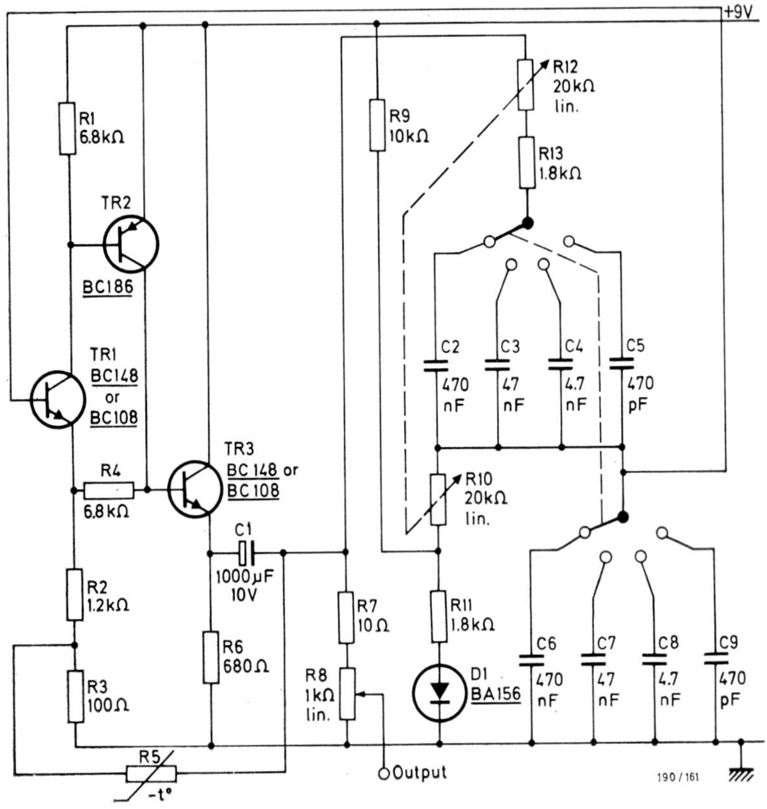

Fig. 161—A.F. signal generator circuit

OSCILLOSCOPE

The reader is referred to the following sources for design information on a simple oscilloscope and auxiliary circuitry:

"The Wireless World Oscilloscope", available from Wireless World, Dorset House, Stamford Street, London, SE1 9LU;

"Oscilloscope Trace Doubler" by W. Kemp, published in The Radio Constructor, Vol. 22, No. 4, Nov. 1968;

"Wide-band Oscilloscope Amplifier" by G. Sowersby, published in The Radio Constructor, Vol. 22, No. 5, Dec. 1968.

The last two are available from Data Publications Limited, 57 Maida Vale, London, W.9.

TRANSISTOR TESTER

The transistor tester shown in Fig. 162 uses a 1mA meter with suitable shunts to measure the collector current of the transistor under test. The base current of the transistor may be switched to zero for measurement of collector-emitter reverse leakage current, I_{CEO}, or to 10μA, 30μA, 100μA, 300μA or 1mA for measurement of static forward current transfer ratio, h_{FE}.

Circuit Description

A meter with 1mA f.s.d. is used and R_9 is chosen to give a total resistance of 450Ω. Switch S_A selects base resistors R_1 to R_7, whose common connection is taken to the collector to avoid the necessity for an additional pole on n-p-n/p-n-p selector switch S_B. Shunt resistors R_{10} to R_{16} are switched in by S_C to give collector current ranges 0 to 1mA, 0 to 3mA, 0 to 10mA, 0 to 30mA and 0 to 100mA. Switch S_D and resistors R_{17} and R_{18} provide a battery check facility. Diode D_1 provides meter protection, limiting the current through the meter to 2mA.

Switches S_D and S_E should be spring-loaded to the 'off' position to prevent current drain from the battery when no measurement is being made. Switches S_A and S_B should be break-before-make, and switch S_C should be make-before-break.

Measurement of I_{CEO} and h_{FE}

Switch S_A is set to position 1, S_B to the correct polarity and S_C to a high range. The transistor is inserted into the socket and with S_E depressed to the 'on' position, S_C is turned until a reading is obtained. For germanium transistors this current, I_{CEO}, should be about 0·5 to 2mA, although for large transistors it may be higher. For silicon types, I_{CEO} should be negligible. If a high reading is obtained the device is probably faulty (short-circuited collector to emitter).

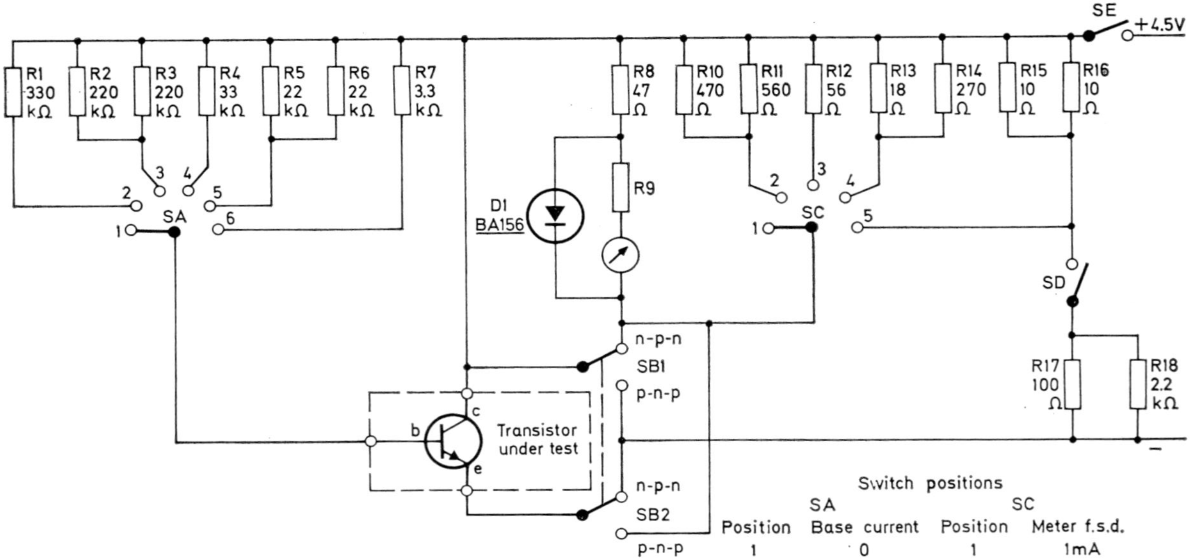

Switch positions

SA Position	SA Base current	SC Position	SC Meter f.s.d.
1	0	1	1mA
2	10μA	2	3mA
3	30μA	3	10mA
4	100μA	4	30mA
5	300μA	5	100mA
6	1 mA		

190/162

Fig. 162—Transistor tester circuit

To measure gain, switch S_A is turned to the position corresponding to the required base current, S_C is set to a high range and S_E is again depressed. If the collector current obtained is I_C and the base current is I_B, the gain of the transistor is given by:

$$h_{FE}=\frac{I_C-I_{CEO}}{I_B}.$$

If I_{CEO} is small enough to be neglected, h_{FE} can be read directly on the meter, but only for base currents of 10µA, 100µA and 1mA. For other values of base current the gain must be computed using the formula given above.

If no increase in collector current is obtained when the base current is switched in, it may be assumed that the transistor is open-circuit.

Battery Check

The internal battery should be checked periodically. With no transistor in the socket, S_C is switched to the 100mA position and S_D and S_E are depressed. A current of 45mA is drawn from a 4·5V battery, so the check is made on full load.

Accuracy

The accuracy of the circuit is determined by the resistor tolerance—5% components are recommended—and the battery voltage. Accuracy also depends on the type of transistor under test and its gain.

The base resistors are returned to the collector of the transistor, and with this arrangement the base current is dependent to a small degree on the voltage drop in the meter circuit. In addition, with germanium transistors the base current will be higher than the value selected, due to the low value of V_{BE}.

With switch S_A in the 10µA position, for example, for a silicon transistor the base current will be 10µA with the meter reading at full scale. With the meter reading at one-third of full scale the base current will be 11µA. A similar error will occur on all ranges. For a germanium transistor the corresponding base current with meter reading at full scale will be 11·5µA, and 12·5µA at one-third of full scale.

For most purposes the error for silicon devices may be neglected. For germanium devices it should be remembered that the base current is approximately 20 percent higher than the selected value.

BALANCE METER

A simple meter circuit which may be used to balance the two outputs in a stereo system is shown in Fig. 163.

Components C_1 and C_2 damp the meter movement. Resistor R_3 is selected to suit the current rating of the centre-zero meter and the amplifier output (typically 10kΩ for a 1mA movement). The voltage rating of the electrolytic capacitors should be half the supply voltage of the system with which the meter is used.

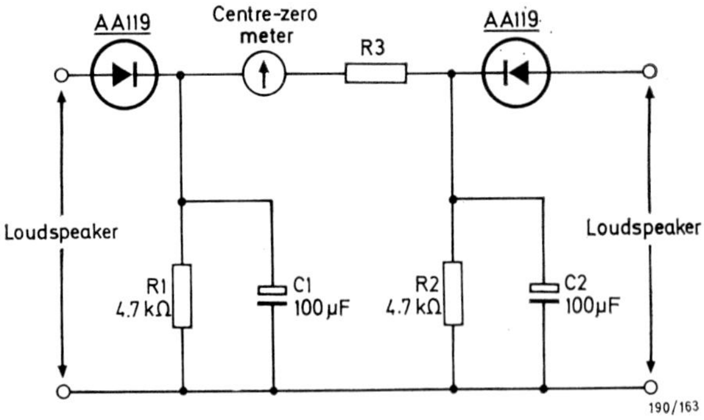

Fig. 163—Balance meter circuit

PREVENTION OF EARTH-LOOP CURRENTS

Transistors may be destroyed by earth-loop currents if the chassis of test equipment is not earthed. The chassis may attain an a.c. potential of 50 to 100V through electrostatic coupling between the primary winding and the core of the mains transformer, as illustrated in Fig. 164a. If, for example, the equipment under test is connected to the output terminals of an unearthed signal generator, this electrostatic potential is transferred to the chassis of the equipment as shown in Fig. 164b. An isolating capacitor in the test leads will not prevent the transfer of the potential because this is an electrostatic effect. If an earthed soldering iron were now applied to the base of a transistor in the circuit, a relatively large current would flow to earth through the transistor and so destroy it.

The formation of this electrostatic potential can be prevented by earthing the chassis of the equipment and test instrument. An isolating capacitor, usually 1μF, should be connected in the lead between a signal generator and the equipment to prevent the flow of direct circulating currents.

Transistors can also be damaged by the use of an unearthed soldering iron. It is possible for the bit, through a breakdown of the insulation between element and bit, to attain a potential of a few volts. When the iron is used on a circuit the potential of the bit may drive a relatively large current through the transistors. Damage due to the defect can be prevented by ensuring that the soldering iron is earthed. However, it should be remembered that when an earthed soldering iron is used the power supply to the circuit should be switched off during soldering.

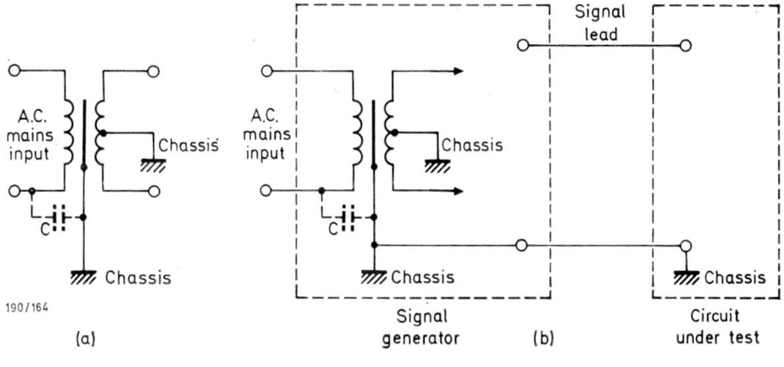

Fig. 164—Diagram showing (a) how the potential of an unearthed chassis is raised and (b) how the potential is transferred to other unearthed chassis

COMPONENT LISTS FOR CIRCUITS IN CHAPTER 10

Audio-frequency Signal Generator (page 254)

Transistors and Diode

Circuit reference	Mullard type
TR_1	BC148 or BC108
TR_2	BC186
TR_3	BC148 or BC108
D_1	BA156

Resistors

Mullard CR25 series (0·3W) ±5% except where stated

Circuit reference	Value	
R_1	6·8kΩ	
R_2	1·2kΩ	
R_3	100Ω	
R_4	6·8kΩ	
R_5 thermistor STC type R53		
R_6	680Ω	
R_7	10Ω	
R_8 linear potentiometer	1kΩ	
R_9	10kΩ	
R_{10} linear potentiometer	20kΩ	matched to within 2dB (standard stereo matching)
R_{12} linear potentiometer	20kΩ	
R_{11}	1·8kΩ	
R_{13}	1·8kΩ	

Capacitors

Circuit reference	Value	Description	Mullard type
C_1	1000μF	electrolytic 10V	C437AR/D1000
C_2	470nF	metallised foil	C280AE/A470K
C_3	47nF	metallised foil	C280AE/P47K
C_4	4·7nF	metallised foil	
C_5	470pF	metallised foil	
C_6	470nF	metallised foil	C280AE/A470K
C_7	47nF	metallised foil	C280AE/P47K
C_8	4·7nF	metallised foil	
C_9	470pF	metallised foil	

Transistor Tester (page 256)

Diode

Circuit reference	Mullard type
D_1	BA156

Resistors

Mullard CR25 series (0·3W) ±5%

Circuit reference	Value	Circuit reference	Value
R_1	330kΩ	R_{10}	470Ω
R_2	220kΩ	R_{11}	560Ω
R_3	220kΩ	R_{12}	56Ω
R_4	33kΩ	R_{13}	18Ω
R_5	22kΩ	R_{14}	270Ω
R_6	22kΩ	R_{15}	10Ω
R_7	3·3kΩ	R_{16}	10Ω
R_8	47Ω	R_{17}	100Ω
R_9	see text	R_{18}	2·2kΩ

Meter

Suitable meters are manufactured by:

Sifam Electrical Instrument Company Limited
Taylor Electrical Instruments Limited

Balance Meter (page 258)

Meter

Suitable meters are manufactured by:

Sifam Electrical Instrument Company Limited
Taylor Electrical Instruments Limited

AVAILABILITY OF ELECTROLYTIC CAPACITORS

The Mullard C426, C431, and C437 series of electrolytic capacitors are being replaced by the 015, 016, and 017 series. Stocks of the older types will be available for some time, but in case of difficulty the newer type can be found from the equivalents list below. No equivalents are given for the C431 series; the nearest value in the newer types should be used.

Old Type	New Type	New Value (μF)	(V)
C426AR/A80	015 12101	100	4
C426AR/A160	016 13151	150	6·3
C426AR/A320	016 12331	330	4
C426AR/A500	016 13471	470	6·3
C426AR/C25	015 14229	22	10
C426AR/C200	016 14221	220	10
C426AR/C320	016 14331	330	10
C426AR/D16	015 15159	15	16
C426AR/D32	015 15339	33	16
C426AR/D125	016 15151	150	16
C426AR/E10	015 16109	10	25
C426AR/E40	016 16479	47	25
C426AR/E125	016 15151	150	16
C426AS/ F1·6	015 90001	1·5	63
C426AR/F6·4	015 17688	6·8	40
C426AR/F12·5	015 17159	15	40
C426AR/F25	015 16229	22	25
C426AR/F80	016 16101	100	25
C426AR/G8	015 17688	6·8	40
C426AR/G16	015 17159	15	40
C426AR/G32	016 17339	33	40
C426AR/G50	016 17479	47	40
C426AR/H2·5	015 18228	2·2	63
C426AR/H5	015 90003	4·7	63
C426AR/H10	016 18109	10	63
C426AR/H32	016 18229	22	63
C437AR/D400	017 14471	470	10
C437AR/D640	017 15681	680	16
C437AR/D1000	017 14102	1000	10
C437AR/E250	016 15221	220	16
C437AR/E400	017 16471	470	25
C437AR/E640	017 15681	680	16
C437AR/F160	016 16151	150	25
C437AR/F250	017 16221	220	25
C437AR/F400	017 16471	470	25
C437AR/F640	017 16681	680	25
C437AR/G100	016 17101	100	40
C437AR/G160	017 17151	150	40

APPENDIX 1

BIASING ARRANGEMENTS FOR H.F. CIRCUITS

Three biasing arrangements used in h.f. circuits—voltage biasing with emitter feedback, current biasing with collector feedback, and simple current biasing —were briefly considered in Chapter 2. The three arrangements are now described in detail, and for each of them an example is worked so that their effectiveness can be compared. In the examples, a transistor with a nominal value of h_{FE} of 100 (h_{FE} spread, 50 to 150) and a nominal V_{BE} of 0·7V (V_{BE} spread, 0·65 to 0·75V) is operating from a supply voltage V_S of 7V; the required emitter current is 1mA.

VOLTAGE BIASING WITH EMITTER FEEDBACK

The circuit of a voltage biasing arrangement is given in Fig. 165. In this circuit the voltage V_B is given by

$$V_B = \frac{R_2 V_S}{R_1 + R_2}$$

provided that the base current does not load the potential divider R_1 and R_2.

The voltage V_E is given by

$$V_E = V_B - V_{BE}$$

and therefore

$$V_E = \frac{R_2 V_S}{R_1 + R_2} - V_{BE}.$$

The operating current, I_E, is therefore given by

$$I_E = \frac{R_2 V_S}{R_E(R_1 + R_2)} - \frac{V_{BE}}{R_E}. \qquad \ldots(1)$$

For the effects of spreads and temperature variations of V_{BE} to be negligible, the voltage V_E must be large compared with the changes in V_{BE}. Furthermore, R_E must also be large if changes in I_E caused by variation of V_S are to be small. In practice, a voltage drop, V_E, up to 3V is required with silicon transistors to prevent disproportionately large changes in operating current with changes in supply voltage, when large variations in V_S have to be accommodated. The corresponding value of V_E for germanium transistors is about 1V.

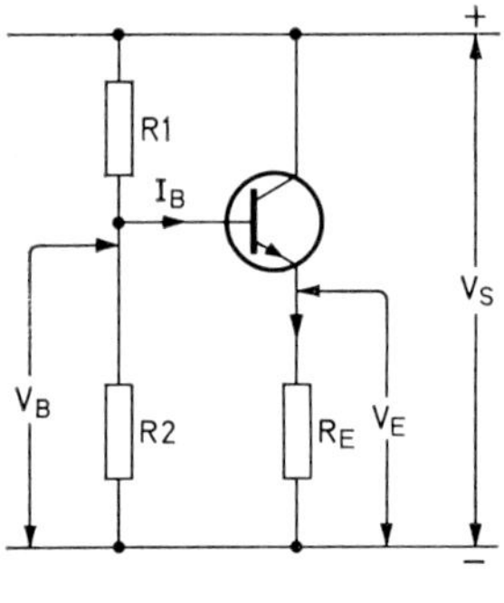

Fig. 165—Voltage biasing with emitter feedback

It has been assumed that the base current does not load the potential divider R_1 and R_2 and, if this is true, the operating current I_E is independent of h_{FE}, and therefore of spreads in h_{FE}. If the voltage across R_2 is significant, however, the operating current is given by

$$I_E = \frac{\dfrac{V_S R_2}{R_1+R_2} - V_{BE}}{\dfrac{R_1 R_2}{(R_1+R_2)(1+h_{FE})} + R_E} \qquad \ldots(2)$$

and the current is then dependent upon h_{FE}.

In the example considered the values of R_1 and R_2 are both 15kΩ and the value of R_E is 2·7kΩ. The nominal collector-emitter voltage is 4·3V, and the actual nominal emitter current is 1·01mA.

Substitution of these values in Eq. 2 gives maximum and minimum values of I_E, resulting from the spread of h_{FE}, of 1·02mA and 0·985mA. Thus, the spread in I_E resulting from the h_{FE} spread is +1%, −1·6%.

The maximum and minimum values of I_E, resulting from the spread in V_{BE}, are 1·025mA and 0·99mA. Thus, the spread in I_E resulting from the spread in V_{BE} is ±1·7%.

If V_S falls from 7V to 3·5V, the nominal emitter current falls from 1·01mA to 0·38mA; that is, I_E falls by 62%.

Eq. 2 reverts to the simple equation (Eq.1) if the following condition is fulfilled

$$R_E \gg \frac{R_1 R_2}{(R_1 + R_2)(1 + h_{FE})}.$$

This equation can be written

$$I_E R_E \gg \frac{R_1 R_2}{(R_1 + R_2)} \times \frac{I_E}{(1 + h_{FE})}$$

or

$$V_E \gg \frac{R_1 R_2 I_B}{(R_1 + R_2)}.$$

Therefore, for the operating current of the circuit shown in Fig. 165 to be substantially independent of h_{FE}, the voltage developed across the emitter resistor must be very much greater than the voltage developed, by the flow of base current, across the d.c. source resistance at the base of the transistor. The d.c. source resistance at the base of the transistor is formed by the parallel combination of R_1 and R_2.

CURRENT BIASING WITH COLLECTOR FEEDBACK

In the current-biased circuit shown in Fig. 166 the voltage V_C is given by

$$V_C = V_S - R_1 I_E .$$

Since

$$I_E = (1 + h_{FE}) I_B$$

then

$$V_C = V_S - I_B R_1 (1 + h_{FE}).$$

The base current I_B is also given by

$$I_B = \frac{V_C - V_{BE}}{R_2} .$$

Hence, if V_C is eliminated, the base current is given by

$$I_B = \frac{V_S - V_{BE}}{R_2 + R_1 (1 + h_{FE})} .$$

Consequently, the operating current is given by

$$I_E = \frac{(V_S - V_{BE}) (1 + h_{FE})}{R_2 + R_1 (1 + h_{FE})} . \qquad ...(3)$$

In the example considered, the value of R_1 is 2·7kΩ and the value of R_2 is 330kΩ. The nominal value of V_{CE} is 4·28V, and the actual nominal emitter current is 1·05mA.

Substitution of these values in Eq. 3 gives maximum and minimum values of emitter current, caused by the spread in h_{FE}, of 1·29mA and 0·635mA. Thus the spread in I_E caused by the spread in h_{FE} is + 23%, − 35%.

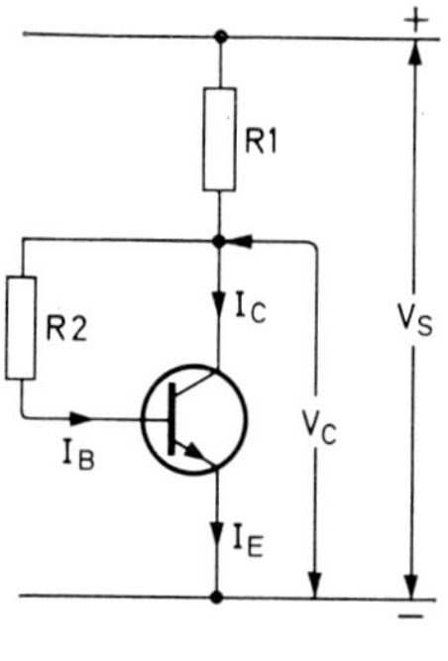

Fig. 166—Current biasing with collector feedback

The variation in I_E with spreads and temperature variations in h_{FE} can be reduced only by applying the following condition:

$$R_1 (1 + h_{FE}) \gg R_2.$$

This lowers the available collector voltage swing and is possible only with large supply voltages.

The maximum and minimum values of emitter current, caused by the spread in V_{BE}, are 1·04mA and 1·06mA. Thus, the spread in I_E caused by the spread in V_{BE} is ±0·95%.

If V_S falls from 7V to 3·5V, the nominal emitter current falls from 1·05mA to 0·47mA, that is, I_E falls by 55%.

In this arrangement there is a spread in the collector voltage resulting from the spread in h_{FE}. In the extreme case, when most of the available supply voltage is developed across R_1 (when the spread in operating current due to the spread in h_{FE} is very low) the spread in collector voltage is directly proportional to the spread in h_{FE}.

SIMPLE CURRENT BIASING

The simplest method of biasing, using only one resistor, is shown in Fig. 167. The current I_B is given by

$$I_B = \frac{V_S - V_{BE}}{R_1}.$$

Thus, the emitter current is given by

$$I_E = \frac{V_S - V_{BE}}{R_1}(1 + h_{FE}). \qquad \ldots(4)$$

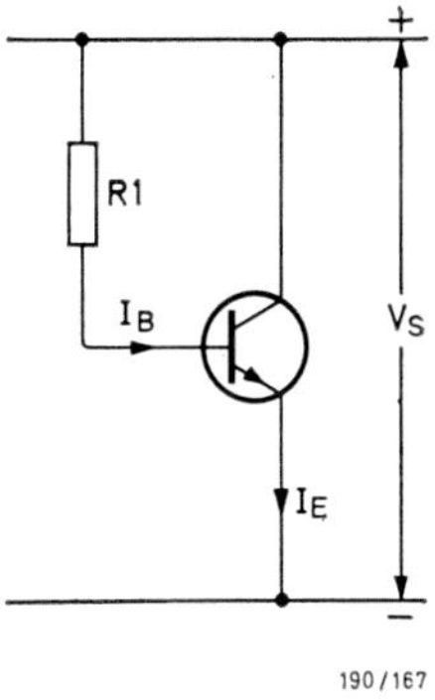

Fig. 167—Simple current biasing

In the example considered the value of R_1 is 680kΩ. The nominal value of V_{CE} is 7V (equal to V_S), and the actual nominal value of I_E is 0·935mA.

Substitution of these values in Eq. 4 gives maximum and minimum values of I_E, resulting from the spread in h_{FE}, of 1·4mA and 0·475mA. Thus the spread in I_E resulting from the spread in h_{FE} is ±49%. Because the emitter current spread resulting from the spread in h_{FE} is high, simple current biasing is suitable only for transistors with narrow h_{FE} spreads.

The maximum and minimum values of I_E resulting from the spread in V_{BE} are 0·944mA and 0·928mA. Thus the spread in I_E resulting from the spread in V_{BE} is ±0·86%.

If V_S falls from 7V to 3·5V, the nominal emitter current falls from 0·935mA to 0·415mA; that is, I_E falls by 56%. However, because there are no resistors in either the collector or the emitter lead, the collector-emitter voltage is equal to the supply voltage. Thus, a higher-value series resistor can be used in the supply line (shared between the r.f. and i.f. stages), if the collector-emitter voltage in this method is to be the same as that in the other two biasing methods discussed. The higher the value of this series resistor the less dependent the emitter current will be on supply voltage.

APPENDIX 2

BBC TEST-TONE TRANSMISSIONS

Details of test-tone transmissions are reproduced in Table 22 from a BBC publication*. The BBC reserves the right to change all or any part of the schedule.

The tones shown in the table are transmitted on Wednesdays and Saturdays. On other days, a 250Hz tone is transmitted in the left channel only from about four minutes after the end of the last programme on Radio Three for twenty minutes. This test may be interrupted from time to time.

The schedule given in the table is subject to variation in accordance with programme requirements and essential transmission tests. The zero level reference corresponds to 40% of the maximum level of modulation applied to either stereophonic channel before pre-emphasis. All tests are transmitted with pre-emphasis.

Periods of tone lasting several minutes are interrupted momentarily at one minute intervals.

---

*The British Broadcasting Corporation, Information Sheet No. 1605 (7), March, 1972.

TABLE 22

BBC Test-tone Transmissions

Test No.	Time	Left Channel (A)	Right Channel (B)
1	T*	250Hz at zero level	440Hz at zero level
2	T+2′	900Hz at +7dB	900Hz at +7dB, antiphase to left channel
3	T+6′	900Hz at +7dB	900Hz at +7dB in phase with left channel
4	T+7′	900Hz at +7dB	No modulation
5	T+8′	No modulation	900Hz at +7dB
6	T+9′ 20″	Tone sequence at −4dB 60Hz 900Hz 5kHz 10kHz This sequence is repeated	No modulation
7	T+10′ 20″	No modulation	Tone sequences as for left channel on Test 6
8	T+11′ 20″	No modulation	No modulation
	T+13′	Reversion to monophonic transmission	

*T is approximately 4 minutes after the end of Radio 3 programmes.

APPENDIX 3

CHARTS AND NOMOGRAMS

This appendix contains miscellaneous charts and nomograms relevant to radio and audio system design:

CHARTS 1, 2 and 3

These three charts are of interest in that they give some indication of the performance required of an audio reproduction system in terms of the characteristics of the human ear and the frequency ranges of musical instruments.

Chart 1 shows equal-loudness contours for the ear, the sound intensity required to produce a sensation of equal loudness at varying frequencies being plotted for various levels of loudness from just audible to the point where the sound is "felt" rather than heard. It will be seen that the ear can respond to all frequencies in the range 15Hz to 20kHz, and to an intensity range of about 120 to 130dB in the region of 3kHz. However, the ear is relatively insensitive to both high and low frequencies in the audio band, especially at low intensity levels. The sensitivity of the ear remains constant with varying intensity above a level of about 80dB and between about 70Hz and 6kHz, but is bass-deficient at lower intensity levels.

Chart 2 relates loudness levels expressed in phons to some every-day noises.

Chart 3 shows the frequency range of a selection of musical instruments, male and female speech. The solid line indicates the actual tone range of the instrument and the dotted line accompanying overtone and action noises.

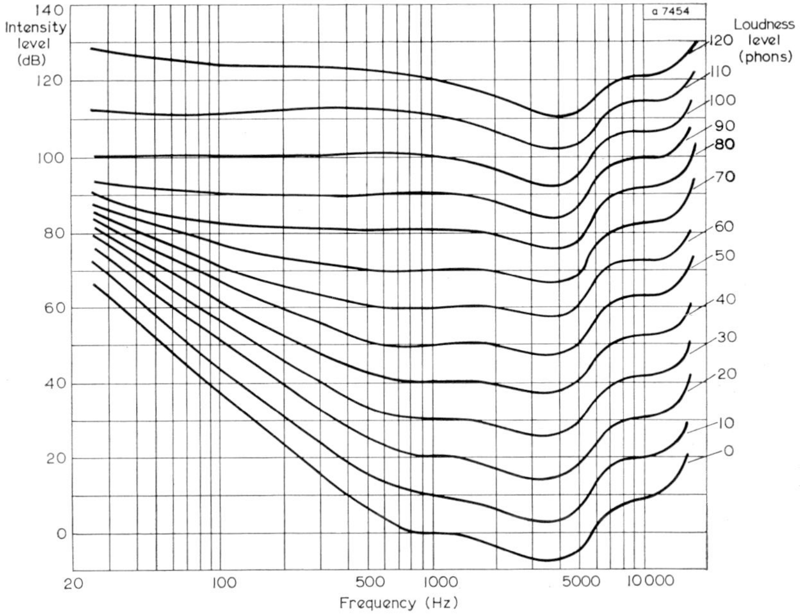

CHART 1. FLETCHER-MUNSON CURVES

Noise	Loudness level in phons
Pneumatic drill	110
Underground train	100
Very loud radio	90
Suburban train	80
Average factory	70
Quiet car	60
Average office	50
Public library	40
Audience noise during concert	30
Whisper	20
Night time in country	10
Soundproof room	approximate zero reference level

CHART 2. EVERY-DAY NOISE LEVELS

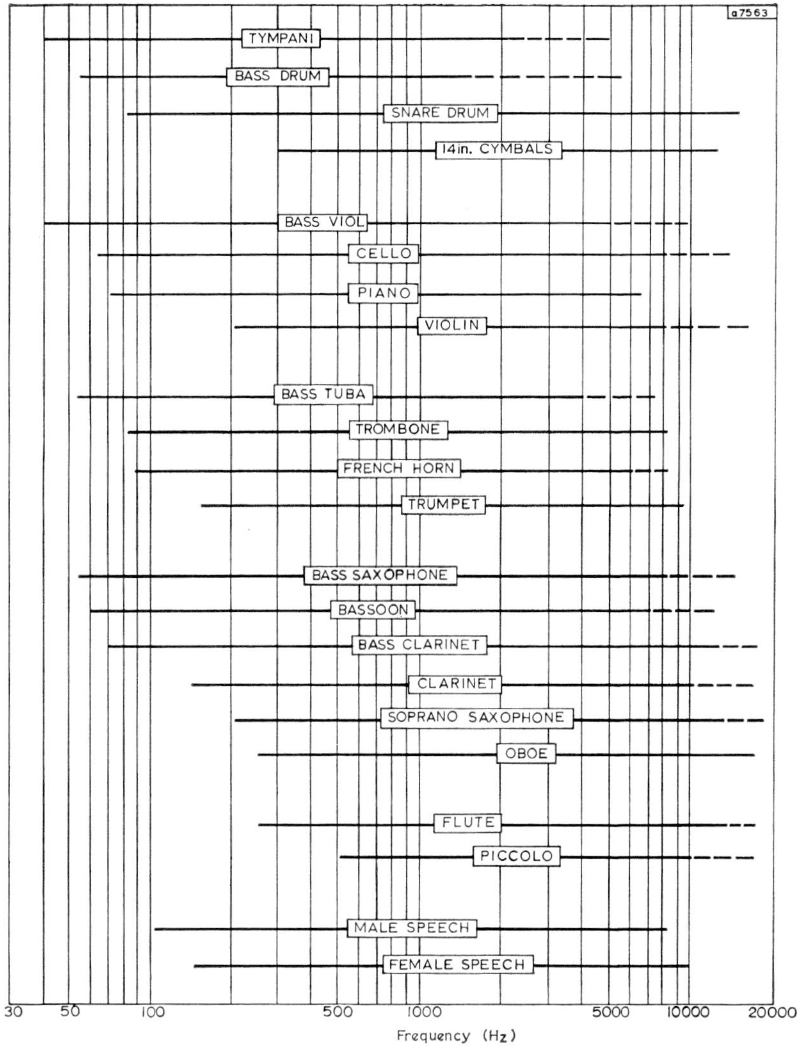

CHART 3. FREQUENCY RANGES OF MUSICAL INSTRUMENTS

CHART 4 RADIO FREQUENCY AND WAVELENGTH

The chart is self-explanatory, values of frequency being tabulated against corresponding values of wave length. The example will illustrate the use of the chart.

Example: What is the wavelength corresponding to a frequency of 70MHz?

Opposite a figure of 7MHz on the scale a wavelength of 42.9 metres is shown. Multiplication by the correct factors for band 5 gives a wavelength of 4.29 metres for a frequency of 70MHz.

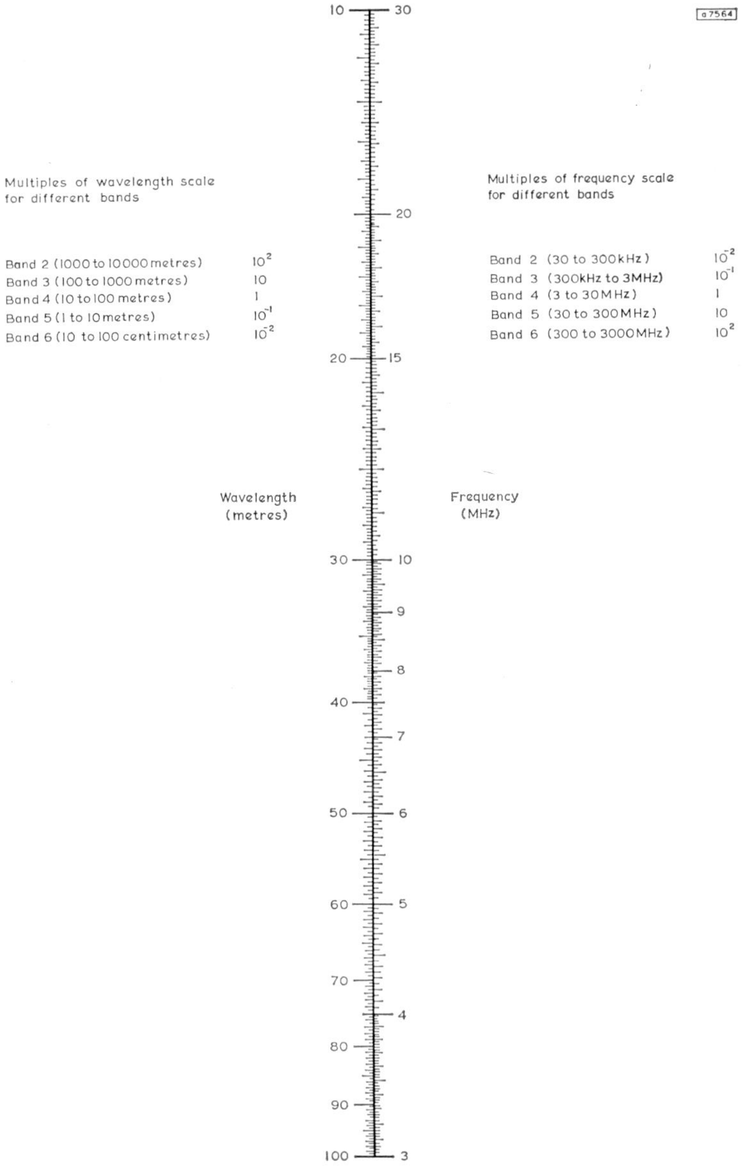

CHART 4. RADIO FREQUENCY AND WAVELENGTH

CHART 5 DECIBELS UP (a) and DOWN (b)

The formula expressing the relationship between power ratio R and decibels dB is:

$$dB = 10 \log_{10} R$$

The formula expressing the relationship between voltage or current ratio, R_e or R_i respectively, and decibels dB, is:

$$dB = 20 \log_{10} R_e$$

$$\text{or} \quad dB = 20 \log_{10} R_i$$

Two points should be remembered when expressing gain in decibels. Firstly, a decibel is a ratio, so gain can only be expressed in decibels when a reference level is understood. Secondly, although voltage and current gains may be expressed in decibels, the decibel is essentially a measure of power ratio. The gain of an amplifier may be given in decibels only if the input and output impedances are equal or some allowance is made for the difference between them. The voltage or current gain of an amplifier is given by the expression:

$$\text{Gain (dB)} = 20 \log_{10} R_e + 10 \log_{10} \frac{Z_1}{Z_2}$$

$$\text{or } 20 \log_{10} R_i + 10 \log_{10} \frac{Z_1}{Z_2}$$

where Z_1 and Z_2 are the input and output impedances of the amplifier.

The use of the charts is illustrated by two examples.

Example 1: What is the power ratio corresponding to 30dB up?

The power ratio is read directly from the 'power' scale (Chart 5a) and is 1000.

Example 2: What is the equivalent in decibels of a voltage ratio of 0·07?

The voltage ratio is located on the 'voltage and current' line (Chart 5b) and the equivalent in dB is read directly as —23dB

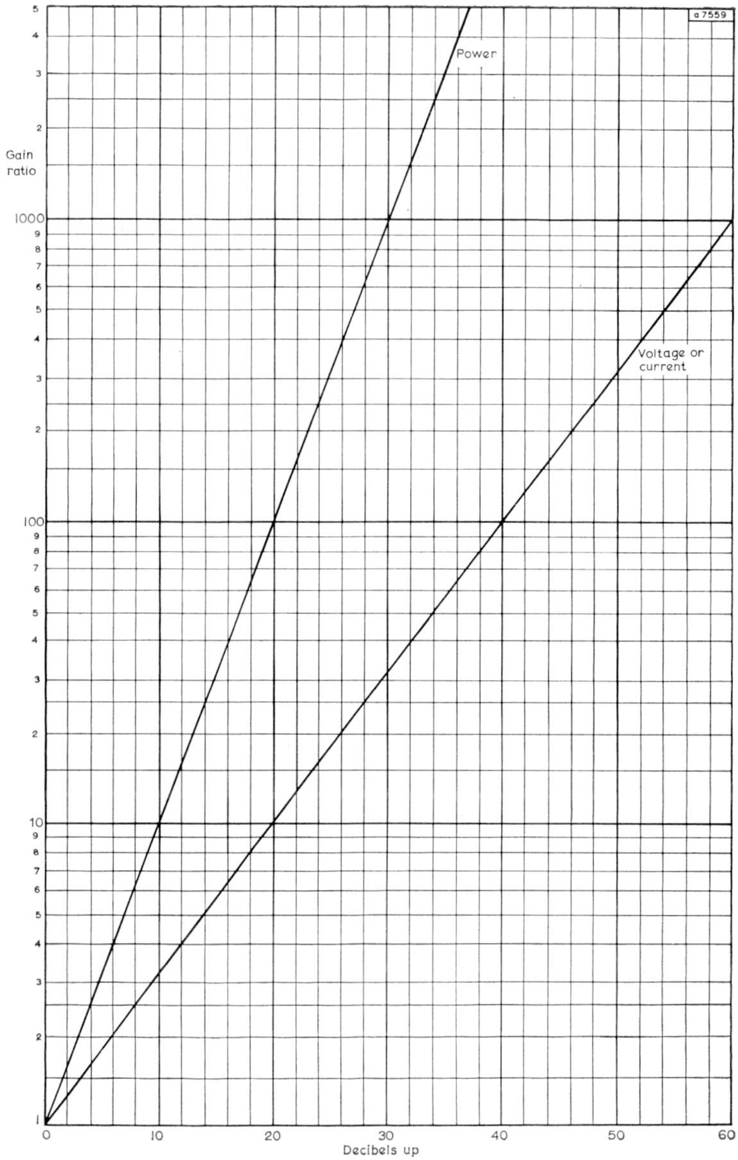

CHART 5a. DECIBELS UP

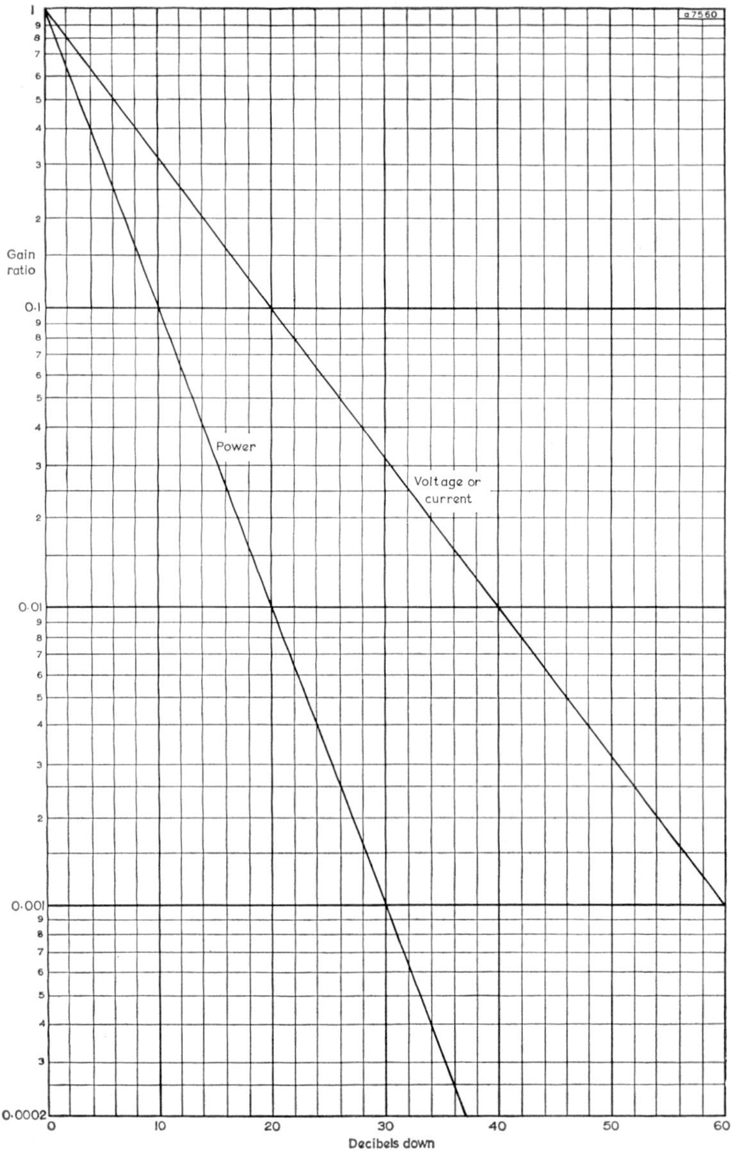

CHART 5b. DECIBELS DOWN

CHART 6 REACTANCE CHART

This chart is used to find the reactance of inductors and capacitors.

Values of inductance are represented by the lines with negative slope; the inductance scale is on the top and left of the chart. Values of capacitance are represented by the lines with positive slope; the capacitance scale is on the top and right of the chart. Values of frequency and reactance are represented by horizontal and vertical lines respectively. The following examples will illustrate the use of the chart.

Example 1: What is the reactance of a 5mH inductor at a frequency of 100kHz?

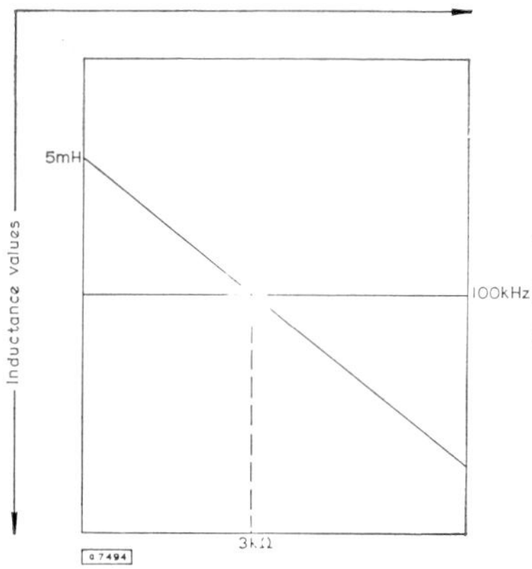

The intersection of the line representing 5mH with the 100kHz horizontal gives a reactance value of 3kΩ.

Example 2: What is the reactance of a 100pF capacitor at a frequency of 40MHz?

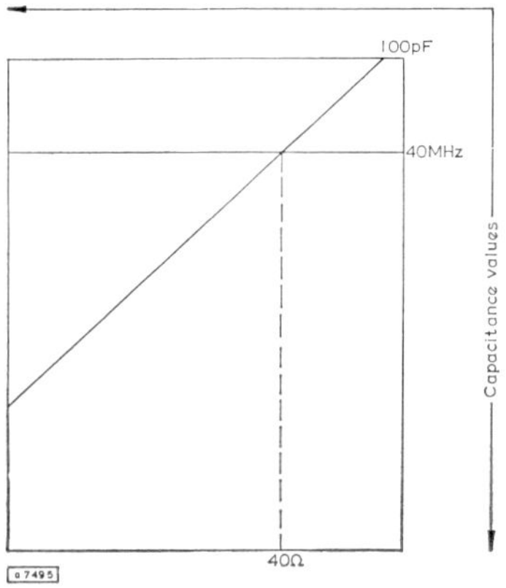

The intersection of the line representing 100pF with the 40MHz horizontal give a reactance value of 40Ω.

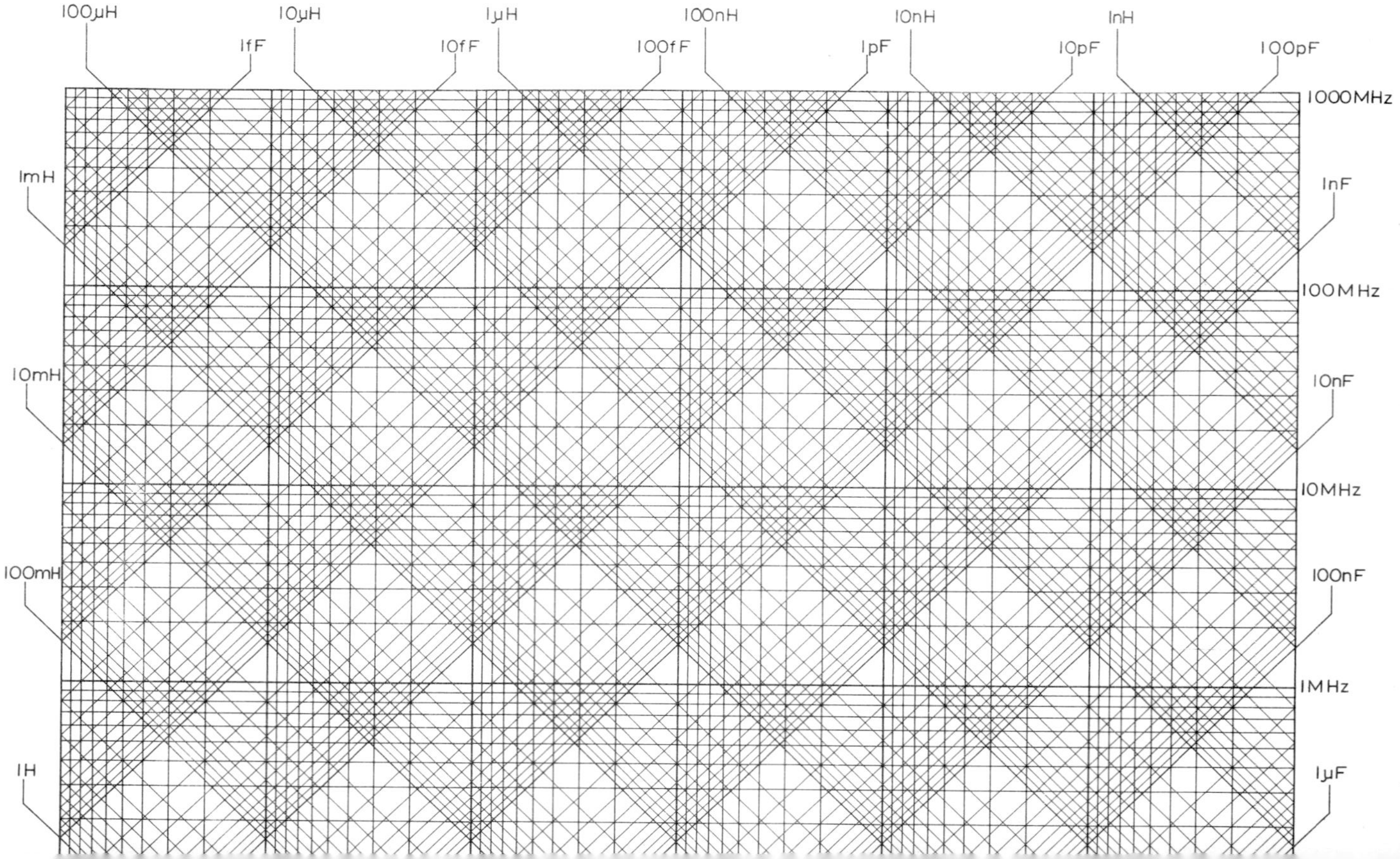
100μH
10μH
1μH
100nH
10nH
1nH
1fF
10fF
100fF
1pF
10pF
100pF
1000MHz
100MHz
10MHz
1MHz
1nF
10nF
100nF
1μF
1mH
10mH
100mH
1H

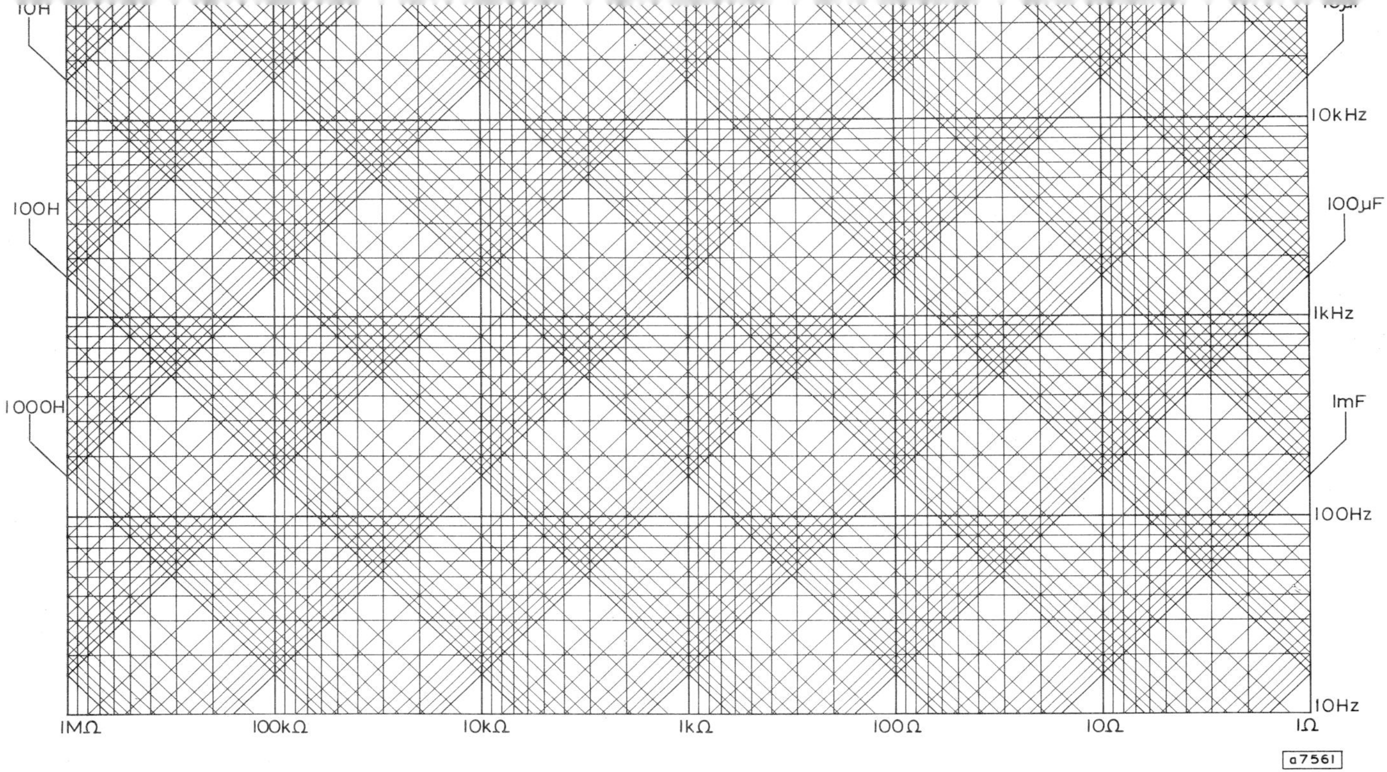

10H
100H
1000H
10kHz
100μF
1kHz
1mF
100Hz
10Hz
1MΩ
100kΩ
10kΩ
1kΩ
100Ω
10Ω
1Ω
a7561